FROM SIBERIAN PRISONER TO DINOSAUR EGG DETECTIVE

LIFE OF THE PAST

Thomas Holtz, editor

MARTIN LOCKLEY with BERNARD SPILKA

FROM SIBERIAN PRISONER TO DINOSAUR EGG DETECTIVE

The Epic Odyssey of Karl Hirsch

INDIANA UNIVERSITY PRESS

This book is a publication of

Indiana University Press
Office of Scholarly Publishing
Herman B Wells Library 350
1320 East 10th Street
Bloomington, Indiana 47405 USA

iupress.indiana.edu

Manufactured in the United States of America

First printing 2024

Library of Congress Cataloging-in-Publication Data

Names: Lockley, Martin G., author. | Spilka, Bernard, 1926- author.
Title: From Siberian prisoner to dinosaur egg detective : the epic odyssey
    of Karl Hirsch / Martin Lockley and Bernard Spilka.
Description: Bloomington, Indiana : Indiana University Press, [2024] | Series: Life of the past
Identifiers: LCCN 2024013144 (print) | LCCN 2024013145 (ebook) | ISBN 9780253070432
    (hardback) | ISBN 9780253070449 (paperback) | ISBN 9780253070456 (ebook)
Subjects: LCSH: Hirsch, Karl F. | Paleontologists—United States—Biography. |
    Dinosaurs—Eggs. | World War, 1939-1945—Campaigns—Eastern Front. | World War,
    1939-1945—Prisoners and prisons, Russian. | Prisoners of war—Germany—Biography. |
    Prisoners of war—Russia (Federation)—Siberia—Biography. | BISAC: NATURE /
    Animals / Dinosaurs & Prehistoric Creatures | SCIENCE / Paleontology
Classification: LCC QE707.H57 L63 2024 (print) | LCC QE707.H57 (ebook) |
    DDC 560.92 [B]—dc23/eng/20240509
LC record available at https://lccn.loc.gov/2024013144
LC ebook record available at https://lccn.loc.gov/2024013145

CONTENTS

PREFACE

This is the life story of Karl Franz Hirsch, a remarkable man who survived two war wounds and Siberian prison camp to emigrate the United States, start a new life, and become one of the world's few experts on fossil eggs. He was, by his twilight years, recognized as one of the leading specialists in the field of dinosaur paleontology. His colleagues and friends, many in various orbits around the University of Colorado, recognized him as a unique, rugged, and larger-than-life character, shaped by an unusual destiny and an ability to overcome some of the more outrageous slings and arrows he had had to fend off.

Karl was born in 1921 on the spring equinox, March 20, near Berlin, and after losing his mother at an early age, lived briefly with a heavy-handed Nazi uncle. Soon his father, a Social Democrat, and his father's new wife moved to Berlin and reestablished the family home. Karl, his sister Anni, and his stepbrother Hans Jürgen passed their school years there during the economic depression and political turmoil that pervaded Germany in the 1920s and 1930s as Adolf Hitler rose to power. Despite the difficult times, and the tough standards of domestic and school discipline, Karl found the time and opportunity to explore the German countryside and surrounding regions, from the shores of the Baltic to Czechoslovakia and Poland, developing a healthy independent spirit. Considering what lay in store for young Karl once he left school, his love of the outdoor life would help sustain his physical and psychic health and his desire for freedom, and eventually make him well suited to paleontological pursuits in remote places.

Upon leaving school in 1938 Karl joined the postal service, but he barely had time to complete his training before being called up for military service.

At the age of nineteen he entered basic training, and within a few all-to-short months he was posted to Romania to launch the great southern front offensive against the Soviet empire (Germans like to say Russia) in Operation Barbarossa. As part of the Sixth Army infantry, he marched more than a thousand miles across Ukraine to the outskirts of Stalingrad, surviving his first Russian winter in Artemovsk where temperatures dropped to 30–40° below zero Fahrenheit. On August 24, 1942, Karl was wounded in his right leg as a Russian tank shell finished off five of the seven men in his platoon. He ran an adrenaline-fueled half marathon to the safety of German lines and was evacuated back to Germany, where he likened the sight of young German nurses to a vision of angels.

After recovering, he went to officer's school, but found he was too independent minded to conform to the discipline and rituals, or to lord it over draftees. (A letter of recommendation describes him as "a clear-thinking, intelligent man whose mind was not swayed by slogans and propaganda.")[1] So he made sergeant and remained a foot soldier with the "fellows" in the trenches. After his first wound, he completed two short tours of easy duty, one in France and a second in Italy where he and his comrades enjoyed the spoils of war—mostly cigarettes, wine, and female company. He was returned in 1943 to the Eastern Front, now in central Ukraine. While leading a unit through dense fog, he stumbled upon a Soviet machine gun emplacement and took the brunt of the fire, again in his legs, while his men escaped unscathed. These wounds were far more severe than those he had previously suffered, and he was lucky to escape with his life and without losing his leg. Following a longer period of recuperation close to his home in Berlin, Karl was posted, in 1944, to engineering duty, building defenses in Germany and Poland. Here, where the war had begun, he was spun from his years of centrifugal military skirmishing, in the more peripheral theaters of war, into the darker centripetal, ever-decreasing and brutal circles of defeat and desperation visited on the Axis powers. He was captured by the Soviets in Danzig and sent to a prisoner-of-war camp in Siberia. Although the war "ended" in 1945, he was not released until 1947, by which time, like fellow survivors, he was in a weak and emaciated condition, a mere skeleton weighing only ninety-four pounds (forty-three kilograms).

Between 1947 and 1952 Karl lived in besieged Berlin, and with his sights set on a better life in the West, he worked for various youth and refugee organizations. During this time, he met Hildegard. They married, and vowed to emigrate and start a new life. This they did, arriving in the United States during the McCarthy era when ironically Germans, ostensibly enemies, were in some quarters more welcome than Soviet communists, who had recently been allies of the Allied powers. After living for a few years in Pennsylvania and Ohio, the couple moved to Colorado where, after thorough vetting by the FBI, Karl went to work as a skilled machinist in a nuclear weapons plant, a choice his friends would sometimes question. During their spare time he and Hildegard hearkened to the call of the great outdoors, and nights under stars and canvas, to become keen rockhounds and build up well-curated fossil collections. It was a field in which amateurs and professionals rubbed shoulders in happy, healthy camaraderie, where a keen eye or good luck could lead to memorable discoveries. Strangely, the prairies of Wyoming, Kansas, and Colorado would remind Karl of the Ukrainian steppe. It was in these wide-open spaces that he would tell some of his stories to fellow "rockhound" Bernie, who would much later help compile this biography.

When he and Hildegard found a fossil egg, they could find no one who could tell them anything about it or provide them with informative leads or literature. The truth was that little was known about the subject, and they were told by professional paleontologists that if they wanted to learn anything about it they would have to "figure it out for themselves." Taking this directive quite literally, Karl set about this task with organized determination. Using his machinist skills to prepare light and electron microscope slides and samples, Karl methodically went about describing and photographing all the interesting fossil eggs he could find and delving into their microscopic microstructure. Happily, he found that much of the specialist seminal literature was written in German, giving him a mildly nostalgic connection with his roots, and at least a modest advantage over some of his monolingual colleagues. Patient and determined from the outset, he began to make his mark as one of a handful of pioneers in the rigorous scientific study of fossil eggs. Enlisting the help of professional paleontologists from the nearby campus of the University of Colorado, he began, as a young student might, to fill in the formal gaps in his scientific education and forge increasingly strong friendships and associations in the world of amateur and professional paleontology. Hearkening back to their days working with wayward refugees in Berlin, the couple, who discovered they could not have children of their own, adopted two recalcitrant boys, with very difficult backgrounds, who proved, to put it mildly, beyond reform. In contrast, their much-loved dogs were models of family obedience and affection.

In 1979, at the age of fifty-eight, Karl published his first scientific paper on fossil eggs. A year later he had a heart attack, but "knew [he] would be alright."[2] In the late 1970s and 1980s the study of eggs, nests, and baby dinosaurs underwent a renaissance, thanks to the widely publicized studies of Jack Horner at the Museum of the Rockies in Montana. While Jack concentrated on the distribution of nests, babies, and questions concerning the ecology and behavior of dinosaurs in nest colonies, Karl knuckled down to the more difficult, less glamorous, and less publicized problems of describing the actual eggs and their complex shell microstructure, and comparing them with the steady stream of new material that began to be reported from all around the world. In 1984 Karl's wife Hildegard Hirsch died of cancer. In his written notes he sometimes referred to her as HH. They had been through more than thirty years of adventurous upheaval, emigration, and marriage. It was a blow, and Karl, now in his sixties, struggled with loneliness and the upwelling of dark memories, but he obeyed Hildegard's "orders" to remain positive and get on with life and his egg research, and to look to his friends for companionship and solace. At times he took Hildegard's directive almost too literally. As he became a recognized expert and mentor to a new generation of students, many who were the age of daughters he never had, he threw himself impatiently into his work, driving sometimes literally at high speed and in high gear into the forefront of research at a time when dinosaur paleontology was enjoying a golden renaissance. For a man who had lost some of his best years, there was little time to lose or squander.

By the early 1990s, Karl had become universally acknowledged as a world expert (eggspert!) on the subject, and among the paleontological fraternity he was widely recognized and respected for his ground-breaking contributions in this unique and specialized field, and to the conference circuits it generated. This well-earned recognition was formalized when Karl became the recipient of the prestigious Strimple Award, bestowed by the Paleontological Society, in 1990.[3] In the same year he was awarded an Honorary Doctoral Degree by the University of Colorado.[4] It is all the more remarkable to consider that Karl achieved this scientific success during a decade when he suffered the loss of his wife, open heart surgery, and a stroke.

Karl is seen posing with his beloved eggs in this publicity photo from the 1990s, the decade in which he was showered with academic acclaim. University of Colorado Museum archives.

So, in the mid-1980s, not long after Karl's wife Hildegard had died, he met fellow paleontologist Martin, who would become his co-biographer. Martin, following a transition in his own life, lived under Karl's roof for almost a year and benefited from having a steadfast friend who had been through much worse than Martin could have imagined. It was comforting and companionable, perhaps for both, to agree upon a modest rent, eat simple food, make field excursions, and sometimes talk as friends about personal feelings or problems. We simply "got on with our work," and often communicated through paleontology. Our work, though different, was similar in that we both studied what are called "trace fossils," including eggs, nests, and tracks—a little outside the mainstream interests of paleontologists, who primarily study skeletal remains. But with many mutual friends and colleagues we read and edited each other's draft manuscripts, and those sent to us by others. Half unknowingly, we were sharing the common experience of working in newly explored areas of paleontology, often with international colleagues who were also new to such fringe disciplines. In such endeavors one could discern the emerging conventions of a new science and appreciate the challenges faced in establishing the study of fossil eggs as a respectable and rigorous science.

Sometimes Karl would excitedly get off the phone with a German friend and, as if he had taken a shot of some stimulant (probably his beloved strong black coffee, good for his low blood pressure), talk enthusiastically in German, before he caught on and switched to English. During this time, Martin never consciously thought of writing Karl's biography, but nevertheless grew to appreciate the contribution Karl was making to this new field of paleontology. As Karl grew older, friends casually suggested that his story was worth telling, and he had even been asked to commit his reminiscences to tape. As a locally well-known self-made scientist, he would talk to school groups and write down some of his memories in a short autobiographical document he called "A Mile in My Moccasins,"[5] a title with some significance for a man who had walked thousands of miles with a few footwear stories to tell. The memoire was never formally published.

One yardstick by which to judge whether a biography is worth writing is perhaps the simple fact that people asked to hear Karl's story. There are also measurable scales of achievement, and as lead biographer Martin was happy to be a part of the group that nominated Karl for the awards and recognition he had so determinedly earned. There are also the less tangible measures of character and friendship. One of Karl's endearing characteristics was that he had not reached the status of respected scientist by traditional paths. Rather, he had ridden a rough road of destiny that fit his strong, rugged, and individualistic character.

While browsing the biography section of a bookshop in England, Martin thought, "Why not write Karl's biography?" The seed was sown, and when he asked for Karl's consent, it was given. It was not an honor or a recognition that Karl sought, nor would he live to see his "authorized" biography completed. It is, however, a story with a posthumous punchline. Befitting Karl's involvement with the paleontological community centered on the University of Colorado, in a pre-Google generation, the US National Science Foundation would honor the memory of a man who had fought on the "wrong side" in World War II, and who had been taken in as a near-destitute immigrant, by providing a quarter-million-dollar grant to preserve his eggshell research collection in perpetuity. For the short bio, the reader can Google the late Karl F. Hirsh and he will grin back at you from the pages of the University of Colorado website. Or, for a story embellished by his own memories and those of friends and colleagues, here is a more complete biography.

A few years after Martin had written drafts of the first few chapters, Karl told a mutual colleague, Ken Carpenter, that "Martin will never finish my biography." Considering how long it has taken, Karl's skepticism about said biographer's finishing power was perhaps well-justified. But a delivery date had never been promised. When Karl died in 1996, the project had been incubating for "only" eight short years! By that time Ken had taken an interest in *Dinosaur Eggs and Babies*[6] and had worked with Karl on a seminal book with this exact title, which was published in 1994, only two years before Karl died. One might say that the science in which Karl had been a pioneer was heading for the mainstream of paleontology. It was never Martin's intention to write scientifically about eggs, as Ken went on to do with another important book, *Eggs, Nests, and Baby Dinosaurs*, published in 1999.[7] Rather, he had wanted to write about Karl the man as a friend, a fellow ex-pat European who had overcome a life of adversity and gone on to make a significant contribution to science. Our discussion of the events surrounding the war never look on a partisan tone. Indeed, Karl was much loved by the community for his lack of any nationalist sentiments, and for putting the horrors of war behind him. The elevation of Karl's collection to the ranks of nationally funded and internationally important components of our scientific heritage, no small tribute to a reluctant prisoner of war, was completed posthumously in 2011 on the ninetieth anniversary of his birth, and the last scientific paper to bear his name and the fruits of his work appeared in 2012. Excuse the delay, but some things take time to incubate and fledge. So, "Hey, Karl, the project was finally completed in time to celebrate the recent passing of your centenary year: 2021."

Now for the "footnote" from Martin and Bernie, perhaps "inevitable" considering that both biographers are retired academics. The devices used to bring forth Karl's voice include his own brief writings, his reminiscence tapes, direct quotes transcribed from Martin's interviews, and a few newspaper and magazine articles. These sometimes tell the same story three or four

different ways, but they are by and large consistent and help authenticate the
chronological and factual background of Karl's epic east-west odyssey. We
place Karl's own words in quotes as often as necessary. In order to distinguish our various experiences, we have taken a page from Theodore White's
In Search of History in which the author refers to himself in the third person,
as "White." In the main text we occasionally use this device, on a first-name
basis, to refer to our personal experiences with Karl.

Martin Lockley and Bernie Spilka, Denver, Colorado, December 2020

Notes

1. Letter of recommendation from Karl Hirsch in the University of Colorado
archives. The letter containing this quote also refers to Karl's very "upstanding character and his brave openness . . . even to his superiors." It was written by Professor
Erich Jenisch of the University of Wurzburg on January 1, 1948. The letter contains
many insights into Karl's character, noting his refusal to be promoted to officer due
to his "firmness of attitude" and refusal to "alter his convictions . . . stubbornly faithful to them to the point of recklessness."

2. After settling in Colorado, Karl Hirsh was often asked to tell his life story.
There are three versions. The first version is a short unpublished manuscript entitled
A Mile in My Moccasins (AMIMM), a copy of which we inherited. The second version is taped interviews that we can call the Hirsh Audio Archive (HAA), which was
converted from the original mini cassette tapes to CD and written transcripts. Evidently these offer posterity the only audio of his reminiscences. The third version is
the messy handwritten notes that the first author compiled during face-to-face interviews. While each of these three sources contains a few unique anecdotes, they often
tell the same story in different words. However, they all represent Karl's authentic
voice, which we often share as "quotations" with the necessary endnotes.

3. See appendix C.

4. See appendix B.

5. *A Mile in My Moccasins*. See note 2.

6. K. Carpenter, K. F. Hirsch, and J. R. Horner, eds., *Dinosaur Eggs and Babies*
(Cambridge: Cambridge University Press, 1994).

7. K. Carpenter, *Eggs, Nests, and Baby Dinosaurs* (Bloomington: Indiana University Press, 1999).

ACKNOWLEDGMENTS

It is to Karl's many friends, and especially Karl's "niece" Ingrid Steinhauer and her family, that we owe a hearty vote of thanks. Without the help of Ingrid and her parents, we would not have had access to many of the original photos they inherited from Karl (unfortunately lost after having mercifully been copied for the university archives). We hope to have done justice to the contributions of Karl's paleontological colleagues by dedicating this book to them and including their input in the text appropriately. With ladies first, we thank Emily Bray, a University of Colorado colleague, for her dual role of being Karl's student in matters paleontological and his teacher in therapeutic matters of the heart. Coauthoring six papers with Karl, including several that were in preparation at the time of his death, constitutes a noble continuation of the egg research tradition, and a bringing of smooth closure to the Karl legacy with only a few tears on both sides, mostly from braving Montana windstorms! Karen Chin, also a University of Colorado colleague, is thanked for her successful efforts to secure funding for the Karl Hirsch egg collection, and the associated, and never easy, challenges of grant administration and reporting. Harley Armstrong, who formerly worked with Karl on Jurassic eggs from western Colorado, has helped in many ways over the years, not least with supporting Karl. Next, much kudos is due Ken Carpenter, formerly with the University of Colorado and the Denver Museum of Nature and Science, for producing two seminal books, one coedited with Karl, which put the study of eggs, nests, and baby dinosaurs solidly on the international map. We thank him too for sending us written memories of his interactions with Karl. We fondly remember the late Judith Harris, who was both empathetic "sister" and professional mentor to Karl, who was, at times, quite literally her student and she his University of Colorado professor for a class he would audit. She, with Emily's help, rescued and helped organize and curate his collection after he died. We thank Konstantin Mikhailov, Russian Academy of Sciences, for his movingly written and psychologically astute reminiscences of meetings and excursions with Karl in Britain, Germany, and the Wild West USA. When you knew Karl, it was all mere history: dice rolls of destiny had thrown us all—friends, colleagues, and biographers (German, Russian, American, British, and Jewish)—together for a memorable melding of paleontological adventure.

PART 1

Headlong into History

CHAPTER ONE

Motherless Child

Young Karl was petrified at the approach of the unknown assailant. Helpless in the dark, and bound up in a dusty flour sack, he could neither see his attacker nor run away. The basement had become a terrifying horror chamber, a medieval dungeon. The blows came senselessly and for no known reason. He had done nothing to deserve such inhumane treatment, and even his confused and terrified young mind knew that this was not right. Who was this shadowy sinister figure looming over him? Did it make sense to believe that mean Uncle "Menne" was inflicting this grim punishment, or had some terrible monster just appeared to strike terror into the heart of a defenseless child?

Thankfully, the cruel and unusual treatment came to the attention of Aunt Lita. She was decent and compassionate and soon found a way to rectify this dire situation. Owing to his mother's sudden "disappearance," this was a desolate and difficult time for young Karl. Too young to think clearly through the confusion, he felt awkward and humiliated. A gentler example: on one occasion he ran an errand for Uncle Menne to fetch something from the railroad station, only to realize that he had his short trousers on backward. No one else noticed, of course, but poor Karl felt acute youthful embarrassment, as if he were at the center of a world that had been turned completely upside down and back to front.

Karl, at that time aged four, and his sister Anni, aged six, were already devastated by the loss of their mother, Friedel, who had died of pneumonia. As the years passed, he came to remember her only through photographs. She had come from a mining family, from the town of Tschernitz in the

Toddler Karl with sister Anni. Germany, ca. 1923.

brown-coal country south of Berlin. Karl could remember being fascinated watching her brother, a glassblower, transform red-hot liquid bubbles of silica into bottles, while other assistants and workers cut precise facets into crystal glassware.

After Friedel's death, Karl's father, Franz, a World War I veteran, quite quickly took a second wife and abandoned his former profession as a miller, leaving behind the waterwheel on the creek at Landsberger Warthe, where the family had lived east of the Oder River, the border between the future East Germany and Poland. Inflation was putting an end to the old mills, and to other small businesses, which one by one went into liquidation. Franz's brother, Hermann, continued in the bakery business for a while before the family moved to Berlin.

It was not only Uncle Menne (also known as Uncle Hermann) who administered corporal punishment. Karl's father, like many adults of that generation, could also dish out stern punishment if the occasion demanded. Karl got a taste of this discipline when he was caught gallivanting about on a jam of floating logs, and he did not soon forget the whacking he received over his father's knee, staring at the unappetizing bilge water sloshing around in the bottom of the small boat in which he had been retrieved.

Because Franz had gained something of a reputation for running around with women besides his wife, he temporarily lost custody of his children, an unusual situation in those days when men usually had the advantage in such matters. Karl went to live with his heavy-handed Uncle Hermann (Uncle Menne), and his sister Anni went to live with her Aunt Lita. The separation of brother and sister, and the cruel discipline of Uncle Menne, only worsened the children's pain over Friedel's untimely death, and Aunt Lita knew she had to step in. By some clever maneuvering, Lita kidnapped Karl and brought him to live with her in the sanctuary of a home with a secure and less unpredictable and abusive environment. Aunt Lita would be Karl's favorite aunt until the day she died.

In this same year, 1925, American paleontologists, led by the swashbuckling Roy Chapman Andrews, were collecting the first complete dinosaur eggs in Mongolia.[1] Meanwhile, the young Austrian Adolf Hitler had just been released from jail after serving nine months of a five-year sentence for treason. Hitler had been convicted following the famous Beer Hall Putsch, a failed coup d'état where he had kidnapped the Bavarian provincial leadership at gunpoint and then announced to the assembled crowd at the beer hall that they had joined him in forming a new national government. During his brief incarceration Hitler had dictated *Mein Kampf,* the turgid and rambling story of his "struggle," which ran to 782 pages when published. Upon his release he had been banned from public speaking. The Nazi Party could muster only twenty-seven thousand members, and it seemed doomed to failure. Following Germany's defeat in 1945, *Mein Kampf* itself would become a banned book for the next seventy years.[2]

In the rural setting of Vietz, near Landsberg Warthe, some one hundred kilometers east of Berlin, Karl fell under the watchful eye of Aunt Lita, her husband (also named Uncle Hermann), and several other step-aunts and step-uncles. It was here that he would form some of his earliest childhood memories. Briefly, in the crowded household, he had to share a bed with Aunt Erna, who being twenty years old had a considerable size advantage when it came to thrashing about at night. To prevent him falling out of bed, Karl had the side next to the wall, and he would often be pinned in place. In the great outdoors, Karl remembered picking mushrooms, chopping wood, helping in the vegetable garden, and watching doves preening in the dovecote. Other rural memories included riding on logging carts, stacking the sheaves at haymaking time, roasting potatoes in campfires, and, of course, playing with Anni. Ironically, given the dizzying turns of history that affected the area during the World War II years, the Vietz Landsberg area and the entire Warthe River Valley would by war's end become part of Poland, as would large areas of once-German territory to the east, where Karl would be captured.

It was during his childhood stay at Vietz that Karl saw his father remarried. During the wedding reception, young Karl parked himself out of sight under the table and set about sampling an interesting selection of the dregs of leftover booze while the adults continued their festivities, oblivious to the inebriated condition of the five-year-old youngster in their midst. By the time Karl was discovered, he had partaken of an interesting and quite potent mixed cocktail and was well and truly drunk. It would not be the last time Karl would enjoy a tipple with friends.

The rural interlude in Vietz was short-lived, as was the peace that had reigned there. In 1925 the family moved to Berlin, where his father and stepmother were living. He remembered on at least one occasion being bundled off on a train to meet up with them in the big city. This was Karl's first view of Berlin, as he sat alone staring out into the quiet gloom of a rainy night: the nighttime illumination of the city streets left a deep, almost surrealistic impression on his young mind. As if divided into two parts, the convex road sloped off on both sides so one side appeared dark and dry and the other appeared so streaming wet that he saw it as a canal running parallel to the street. He wondered if this was perhaps a city of canals.

After Karl's quick succession of separations from his mother, father, and sister, the fragmented family began to reunite in Berlin. The year was 1926 and Karl's half-brother Hans Jürgen was born. The family lived first in a simple five-story apartment building on Tanroggener Strasse. The complex was of simple design: a rectangular edifice surrounding an enclosed courtyard with rubbish bins, washing lines, carpet-beating frames, and a few bushes. Despite the urban setting, the complex still had some of the trappings of a bygone rural age, notably the cowshed from which fresh warm milk was delivered daily. The cows also provided manure for the allotments, and on occasion it was Karl's duty to load up and deliver dung mucked out from the shed, using a little red wagon. For the same purpose Karl and Anni also scooped up "horse apples" (*pferdeapfel*) dropped by draft horses that delivered milk and other commodities to the neighborhood.[3] The children would also trundle the family's large accumulator radio battery to the shop on weekends, to get it recharged.

The design of the apartment was simple as well, consisting essentially of two rooms and a kitchen, with a small storage area, pantry, and toilet. The three children shared one room and their parents shared the other, which doubled as a living room during the day. Suffering from subconscious restlessness, Karl would sometimes sleepwalk, carrying a blanket and a pillow from the crowded bed and curling up in the small storage area. The home was heated by a wood stove and a coal-fired tile stove, which had compartments that could be used for keeping food warm. The home was also supplied with both running water and electricity, carried by exposed pipes and wires on the interior walls. Outside, the street lamps were gaslit, and hand pumps were still used for watering horses.

For many a resilient human soul in their formative years, if there is some measure of order in life and some contact and affection among relatives and friends, a little upheaval and chaos is acceptable. Thus, Karl recalled his youth as relatively normal. In the broader context of his socioeconomic world, Karl counted himself as neither rich nor poor, despite the depressed times. He knew to not expect too much. It was simply accepted that there was little or no money for replacing clothing or even a lost pencil. He regarded himself neither as privileged nor as deprived, but he lived in a world where luxuries such as pocket money simply did not exist for fifty weeks of the year. Despite a tendency toward heavy-handedness and the liberal use of corporal punishment, parents, teachers, and relatives cared about order and established traditions. Karl and his siblings were assured of access to decent food, education, and an ordered household, and the youngsters could look forward to treats and modest presents on birthdays and other special occasions. Typical Christmas fare was a book, a bar of chocolate, a plate of nuts, and some fruit.

Karl was not fussy about food, but he never liked chicken skin. Once, when served chicken by his stepmother, he contrived to dispose of the skin under the dresser in his bedroom, where it lay for a while gathering dust. When his stepmother later found it, she mistook it for a dead mouse. When the truth was revealed, Karl's punishment was two stern whacks.

From 1927 to 1931, Karl attended the Volksschule.[4] This elementary school was run with firm standards of strict discipline. Boys and girls sat in separate circles and were not permitted to play together. If you wanted to speak, first you had to raise your hand. During breaks the boys and girls were segregated and marched around the schoolyard in separate circles, as much for compulsory exercise as for recreation. The tramping of endless feet, some shod with hobnail boots, gradually wore a shallow indentation in the school's solid granite steps. With the powers of geologic observation that he developed in his adult years, Karl immediately spotted this enduring example of erosion when he returned to visit his childhood haunts many years later.

Indoors and in wintry weather the children could exercise on ropes, vaulting horses, and other gymnastic equipment. In winter there were opportunities to ice-skate using simple metal blades clamped onto one's shoes. Such primitive equipment was not conducive to the high performance one could attain when wearing proper skating boots. So the children teetered, rocked, and stamped their way around the ice without the benefit of the ankle support provided by today's skates. It was a simpler time, when austerity and makeshift equipment were the norm.

Depending on the season, the school provided milk and cocoa to supplement the sandwich Karl carried to school in his little waist satchel, or *stullentasche*. Young boys wore shorts and girls wore skirts, so in cold weather Karl and the other youngsters also wore long woolen stockings that could be rolled up over the knee. It was considered "sissy," however, for boys to roll up their socks, and so, with youthful bravado and disregard for the freezing temperature, they defiantly rolled them down below their knees.

In this no-nonsense setting the children either enjoyed or endured the same teacher for all four years of their primary schooling. Moreover, their teachers kept in touch with parents to make sure that there were no deviations from the norms of acceptable behavior. This does not mean that there were no opportunities for recreation and high jinks. In the boys' changing rooms there were a series of cubicles furnished with pegs on which to hang garments—and test one's mountaineering skills. One day, an unfortunate youngster slipped and managed to skewer his buttock on one of the metal clothes pegs, and narrowly missed being skewered right in the rectum. This incident made a big impression on Karl and his classmates and made them think twice about scaling the walls of the changing-room facilities. In this case the injuries that might be sustained threatened to be more gruesome than the punishments suffered at the hands of teachers or parents.

One weekend excursion found Karl on an outing to the main aquarium in Berlin. He had somehow procured a fistful of little paper stickers, which he managed to deposit at strategic locations along the way and on the glass of the fish tanks. This rather obvious trailblazing did not go unnoticed, and Karl soon found himself having to retrace his steps and remove all trace of his decorative expedition. On the outskirts of town, where the sewer system fed the apple orchards, Karl sometimes got away with minor infractions. He and his partners in crime could usually climb apple trees and harvest the forbidden fruit without getting caught. It was a time when youngsters still relished contraband apples as exotic treats.

Even so, Karl once suffered further his father's rather severe discipline simply for complaining that he had been wrongly punished at school. His father's logic was that since Karl did not complain when he did something wrong and got away with it, he should be punished again to even the score and teach him not to bellyache about trivial matters. These were the no-nonsense rules of daily life: conform before you complain. Needless to say, Karl has no memory of ever being hugged by his father, or of ever seeing much of an empathetic side to the man's nature.

Karl's father Franz was thoughtful and candid about political matters, however, and stood up for his ideological beliefs. In his large extended family, it is perhaps no surprise that there were overt political differences, sometimes obvious to all. His father was a Social Democrat, but on this same side of the family his Aunt Lenchen and his fearsome Uncle Menne were both Nazis. Karl was aware of such differences from an early age. These divergent ideologies, and especially his father's political affiliation, were to play a significant role in his father's ability to sustain work. They ultimately affected the family's income and their choice of schooling, and they had ramifications for Karl's educational options.

It seems that young Hans Jürgen, as the baby of the family and the blood child of both parents, received a certain amount of preferential treatment. Karl and Anni were generally held to somewhat stricter standards of discipline and behavior, as befitted their greater maturity. This was in part because Hans had suffered a kidney infection and so had to be fed a special diet without any meat. Hans sometimes took advantage of this special treatment, and he tended to be fussy about food throughout his youth. During the war years, when at home recovering from his second wound, Karl remembers his surprise at seeing Hans wolf down a ration of pork chops that Karl had brought home. It was the first time he had seen his brother eat meat, and it was obvious that hunger had overcome his once-fastidious eating habits.

When Hans maliciously interfered with, and spoiled, their careful handwriting (cursive) homework, by shaking the table, neither Karl nor Anni got any sympathy for crying to their parents. Quite the reverse, they got a whacking. On another occasion they found themselves in hot water after they had researched some adult dictionaries and biology books and were suspected of exploring forbidden illustrated sections on sex and human anatomy. It seems that their illicit interest in such adult subjects was betrayed by their inept efforts to replace the books properly on the shelves.

Karl could remember only so much of his primary school years and his home on Tanroggener Strasse. Two young girls from a Catholic family lived in the same building and also attended the Volksschule. Their father was a shady character who bet on the horses and beat his wife and his children when he lost money. Most families in the district were Lutheran, but Karl occasionally attended a service at the small Catholic chapel with his schoolmates. There he was amazed by the different atmosphere and was especially impressed by the priest's ability to produce smoke from the swinging incense burners.

In the early 1930s the family moved from their Tanroggener Strasse apartment, with its inward view of an enclosed courtyard, to Ylsenburger Strasse, where they had procured a larger, three-room residence with an outfacing view of the street. In those days, children were expected to conduct themselves in ways that did not violate any of the accepted community norms of behavior and decency. To this end, the housing complex had its own self-appointed *fräulein*, who made it her business to check on everyone else's business. This imposing so-called pillar of the establishment perched herself at her windowsill and watched to make sure that no youngsters living at the complex who fell within her eagle-eyed field of view were misbehaving, and that boyfriends and girlfriends maintained the appropriate standards of decorum. Fräulein was so much a part of the institution that Karl saw her there again a full generation later when he returned long after the war.

In his tenth year, Karl graduated from the Volksschule and was eligible to move on to the technical Friesen Oberrealschule.[5] These were the depression years that Hitler exploited to rebuild the Nazi Party under the guise of National Socialism. It was common for kids like Karl to see men standing in line for work and long lines of women waiting to buy food. One could often sense the frustration and disgruntlement of the crowd. Sometimes a drunk or an agitator would get rowdy and start throwing cobblestones. This would precipitate the intervention of the fire brigade and their hoses and would also bring out the police, some of whom would retaliate by throwing the cobblestones right back at the demonstrators. It was tempting for Karl and his friends to stop and watch these charged incidents, but they were expected to walk straight home from school or face a whacking—that is, not to linger to watch evidence of adult misbehavior!

In a whirlwind campaign after the October 1929 stock market crash, Hitler promised hope and a future to the masses by pledging a strong Germany, and by stirring up patriotic nationalist resentment against the reparations demanded by the Treaty of Versailles. After having polled only eight hundred ten thousand votes in 1928, the Nazi Party gained six and a half million votes in September of 1930 and was propelled to the status of second-largest party in the Reichstag, or German parliament. Hitler was essentially in power.[6]

Around Europe and across the entire world, thoughtful observers began to watch the rise of Hitler with a sense of foreboding. But many people were blind to the events that were unfolding before them. Like a man in a trance, strangely disconnected from his higher brain functions, Hitler regressed with disconcerting ease into a saurian state, and cast a limbic spell over mesmerized audiences. Underscoring the irrational futility of his calling, Hitler admitted, "I go the way fate has pointed me, like a man walking in his sleep."[7] According to the writer Laurens van der Post, only Carl Jung knew the real truth about Hitler. Van der Post quotes Jung as stating not that Hitler "rules Germany" but that he "is Germany" and is "more of a myth than a man, the loudspeaker that makes audible all the inaudible murmurings of the German soul," a soul that at this point in time was submerging itself in its murky subconscious.

Although Karl at first registered at the Friesen Oberrealschule, he attended only briefly, as his family could not afford the fees and he was soon obliged to withdraw. He then enrolled in the less-prestigious Aufbau Schule.[8] This institutional educational step down was the direct result of his father's declining status as a Social Democrat. When Hitler came to full power, in 1933, Franz Hirsch lost his job as head of the Allgemeine Ortskrankenkasse—a community health insurance office within the Civil Service. In this same year, across the Atlantic in the United States, in Colorado, where thirty years later Karl would make his home and dig for fossils, the "Dirty Thirties" Dust Bowl was just beginning. The outlook was dark and dirty, as monster dust storms killed children, the elderly, and starving livestock. Germans and other immigrants looking for a new life on the High Plains would escape Germany only to face privations almost as dire as those that the upcoming war would visit on the Old World.

Franz was one of those not mesmerized by the Nazi spell. He opposed Hitler and war in general. He had served reluctantly in the first "great war to end all wars" but had participated in only one Zeppelin bombing raid on London. This had been enough offensive action for him. Through a little constructive paperwork, he managed to get his orders adjusted and was transferred away from the front line to a position where his duties involved delivery of supplies. Franz shared his feelings and experience with Karl and advised him to stay away from the front line in time of war. Although Karl later saw more than enough frontline action, there were times he heeded his father's advice in arranging potentially favorable and safe supply-side deployments.

Franz did not take this political sacking lying down, and after fighting the decision in the courts for six months he got his job back. This reinstatement was short-lived, however, and in 1934, by order of the Minister of State, he was laid off again. Once more, Franz fought the political discrimination of the system and was able to obtain a less satisfactory reinstatement at a lower rank. In the climate of the times such injustices could not be talked about openly, but Franz nevertheless instilled in Karl a healthy skepticism toward authority. In the presence of Nazi visitors, Karl was able to characterize his father as a veteran of bombing raids on London, rather than a dissenter with little sympathy for Hitler. The reality was that the family's income was eroded, and Karl's Aunt Lita had to help out by providing extra food during these interludes when his father was made redundant.

While on a visit to Munich in 1932, Karl caught a glimpse of Hitler close up and in the flesh. The future führer had just pulled up at a restaurant in a big black Mercedes and was but a few feet away as he stepped out with a henchman on either side. "If I had known what he was going to put me through," Karl later recalled, "I would have tripped him up right then and there."[9] The eleven-year-old was not then about to change the course of history.

Notes

1. See chapter 14.
2. Hitler's infamous autobiographical book *Mein Kampf*, meaning "my struggle," was published in 1925, with a second volume appearing in 1926. It carried a subtitle

translated as *The National Socialist Movement*. The book was banned in Germany after the war (see chap. 2, note 1).

3. Karl's reminiscences often involved key German words that give flavor to his memories. Where they are used, we provide the appropriate translations in the text, and notes are not necessary unless indicated. See notes 4 and 5. The German names for streets and locations are not translated unless particularly significant or interesting.

4. *Volksschule* was a "compulsory school," and it might be translated into American English as "public school" (not the British sense).

5. *Friesen Oberrealschule* literally translates as "free upper school."

6. The vast literature on World War II is obviously relevant to Karl's biography. We have used the device of boxes to place events in Karl's life in their broader historical context.

7. Laurens van der Post, *Jung and the Story of Our Time* (New York: Vintage, 1976), p. 19.

8. *Aufbau Schule*, "building or construction school," has the connotation of "technical college."

9. See Preface note 2. All direct quotes from Karl Hirsch are taken from his unpublished autobiographical notes and tapes, provided to the authors during his lifetime.

CHAPTER TWO

Wänderlust

In secondary school, the tradition of firm discipline and moral standards continued. The pupils continued to learn in the same classroom with the same teacher, with only a few excursions to other classrooms and little exposure to other teachers. Karl's teacher was Herr Klaetsch, who kept in close contact with the parents of all his pupils, to ensure a high standard of discipline. The students could not participate in school skiing excursions or other recreational activities if they did not maintain acceptable academic performances. On one occasion Karl, who usually made passable grades, was very near a failing grade and had to beg his teacher for the requisite score to qualify for a ski trip. Mr. Klaetsch relented on two conditions: Karl had to put in extra effort as the class progressed, and he had to show himself as a willing and helpful assistant on the forthcoming ski trip. It turns out that he was expected to help by waxing his teacher's skis. Showing little enthusiasm for the bargain, a rebellious Karl waxed one ski for climbing and the other for smooth downhill descent. Herr Klaetsch was destined for a rather erratic ride as one ski grabbed and the other rode smoothly.

The school offered a no-nonsense academic curriculum, with few easy or "fun" classes such as woodworking. The school also operated a program on Saturday mornings that allowed, among other things, the exercise of political and religious expression. Karl was confirmed in the Lutheran Church and qualified to wear long trousers for the first time. Catholics and other denominations were allowed to attend their own assemblies, while others, if they chose to, could participate in youth-group activities. As Hitler rose to power, he introduced the option of participating in the Hitler Youth, later to become the only legal youth organization. From an early age, youngsters

were very aware of politics, and many wore the insignia of their family's affiliation in school, whether it was the Christian cross, the communist hammer
and sickle, the three arrows of the Social Democrats, or later, as other insignia
became outlawed, the Nazi swastika. Such political diversity reflected the tu-
multuous times. People described the revolving door of government politics
as the *witshaft aukurbeln*—in reference to the hand crank used to start cars.

As Karl entered his teens, Anni turned fifteen and left school (and home) to
train as a nurse. She and her stepmother had never had a very close mother-
daughter relationship, so the day-to-day separation was probably for the best.
This situation, however, left Karl with a greater measure of responsibility for
young Hans, who—being five years younger and suffering from a delicate
health condition—was not expected to do much heavy work. So, invariably it
was Karl who brought the coal and potatoes up from the basement. This particular chore always made him uncomfortable. Remembering the treatment
he had received at the hands of Uncle Menne, Karl was again apprehensive
when running such errands to the subterranean world.

On the one hand, in Karl's young psyche, day-to-day Germanic tradition
and discipline bred a conventional mentality that desired order where order
had sometimes been lacking. On the other hand, such tradition and discipline bred a powerful independence and self-reliance that served him well for
the remainder of his years. His father's nonconformist ideological position
also helped him learn to think independently and to display healthy skepticism toward regimented authority.

He soon had the opportunity to exercise both facets of his character when
he joined the labor youth organization Bündische Jugend and found it his
passport to travel and adventure. He and his loose-knit group of friends were
surprisingly free to embark on camping expeditions and hitchhiking excursions around the provinces. Such travels put him in touch with geography
and nature and satisfied his youthful wänderlust. It was in this environment
of campfire camaraderie, often singing favorite Russian and Cossack songs,
that Karl began to exercise independence in the real world, constrained only
by the code of honor necessary to ensure the group's cohesion and its safe
arrival at its destination.

The urge to explore was not born of any military aspirations. As it was in
many an inquisitive youngster and future paleontologist, Karl's wänderlust

gene was strong, and it drove him on to explore the great outdoors and to dream of exotic faraway places. Karl often read of the adventures and exploits of Marco Polo, Lawrence of Arabia, and other heroic explorers. Another favorite writer was Karl May, who created the story of the fictional Apache chief Winnetou and other adventurous Indian heroes on the wild American frontier. Little did young Karl know at that time what the future had in store, or how the youth group's rural excursions and forays might be preparing him for rigorous years of foot-soldiering on the wildest and most dangerous of frontiers.

The Bündische Jugend group roamed the country far and wide. Every April they hitchhiked to the Baltic Sea for the first swim of the season. On their rural ramblings, they slept in haystacks, camped in a tipilike Lapland "kote," and hitched rides on logging carts. In town they visited museums and art galleries and inspected the architecture of bridges and fine buildings. They also employed mild subterfuge to obtain maps. By telling the proprietors of the shell shops that they had been sent by relatives to procure the necessary directions for a family trip, they gained access to these adult, potentially illicit, cartographical materials.

On occasion, the group even crossed freely into Czechoslovakia, without customs inspection. There they bunked in farm buildings during winter ski trips, and hitched rides up the long slopes on horse-drawn sleds, before gliding several kilometers back down. Karl had grown into a tough but wiry youth, who swam, skied, and sledded like any healthy adolescent. He was a capable skier by the age of ten, and in addition to outings on German and Czech snow he skied on Poland's 1,600-meter "giant mountain"—Reisengebirge—a favorite Christmas destination for school excursions.

The youth group was freer and less organized than a school group or a boy scout troop and had no formal uniform, rigorous disciplinary code, or written record of activities. This casual organizational structure was to prove an advantage in later years, when the group was outlawed and its members were investigated by a new breed of Nazi authority.

One of Karl's closest youth-group friends was Klaus. Like Karl, Klaus had a nonconformist father (who belonged to the Communist Party and, as a result, in 1933 was thrown in jail as Hitler came to power). The youth group survived for a few more years as a legal entity but was officially outlawed by authority of the Nazis in 1936. From then on, only the Hitler Youth ("Hitler

Jugend") was recognized as a legitimate youth organization. Because of his family's politics, Klaus's father ended up in a concentration camp by Nazi decree. He was liberated by the Allies in 1945 and was put in charge of the police in Wilmersdorf (an inner-city part of Berlin). Here, again, his political leanings got him in trouble, and the British arrested him for passing information to the Russians.

Karl and his fellow youth-group members continued to associate and travel under the clandestine cloak of friendship, with no formal group structure to give them away. Nevertheless, they were watched and scrutinized. On one occasion Karl was pulled out of school and taken to the police station, where he and several other group members were questioned about Atze, their leader, who happened to be Jewish. Karl and the other group members were encouraged to falsely denounce Atze as a homosexual. They resisted the pressure to betray their friend and denied that they belonged to any organization. In truth, they were no more than a loosely knit, and perfectly normal, group of friends, neighbors, and school acquaintances. Atze later escaped to South America, where he married the daughter of a well-to-do railroad owner.

The freedom of the youth group contrasted with the discipline and politicization of school life. In the early 1930s young secondary-school children were becoming aware of political and religious affiliations in the community. By choice many wore badges denoting their support of various parties. It was a sign of the turbulent economic and political times that schoolchildren wore such insignia. The different flags and banners often adorned apartment buildings and even different sides of Berlin streets, revealing the political rifts often dividing the neighborhoods.

Karl was a capable and generally conscientious student who always gotten a grade three or better on the five-point scale. Grades of four and five were substandard and would almost inevitably lead to some punishment, such as disqualification from the right to participate in school skiing excursions. Despite his habit of not always paying attention in school, and of sitting at the back of the classroom where he could read books hidden under his desk, an activity that teachers today might even consider healthy, Karl was able to avoid poor grades, and the subsequent consequences. During his secondary-school years Karl became quite a keen artist, and little by little he

Hitler's rise to power did not immediately alter school curricula or affect Karl's secondary education. Certain books were banned, and some were even burned by fervent Nazis. But none of the standard school texts came in for vilification, nor were there any revisionist attempts to rewrite world history. In Karl's recollection, the censorship was less blatant and restrictive, and in many ways considerably less paranoid, than the atmosphere he experienced in America, in 1952, when he arrived in the middle of the McCarthy era. By the mid-1930s, however, at certain school assemblies and functions, it became necessary to sing the Hitlerian anthem, "Die Fahne hoch, die Reihen fest geschlossen," as well as the national anthem, "Deutschland, Deutschland über alles."[1] Those, like Karl, who were not strongly persuaded by the Nazi fervor would register their protest by not singing the Nazi hymn or by silently mouthing some rebellious alternative.

accumulated a pictorial record of his travels in Berlin and the countryside beyond. His artistic talents even earned him a 100-mark prize, and four of his colorful paintings from that time survive to this day.[2]

As a teenager, in a technical school, Karl learned to appreciate, love, and draw the symmetry and order of bridges, machinery, and other mechanisms, and he entertained aspirations of becoming a civil engineer. In later life he became accomplished as a machinist, and as a paleontological preparator, photographer, and curator of his own highly organized and scientifically important collection of fossil specimens.[3]

In his teenage years Karl's outdoor, skiing, swimming, and hiking activities developed into a growing interest in competitive sport. For a while he had a job as a ball boy at an exclusive tennis club. It was unusual at that time for schoolboys to find paying jobs, and Karl had to maintain the expected grades in school to qualify for the privilege. Once in the club, he had access to additional income that could be invested in skis, books, and other items that he could not otherwise afford. On one occasion Karl rubbed shoulders with the US player Bill Tilden, whom he remembers for discarding a set of

In 1936 Berlin hosted the Summer Olympics. Hitler, who firmly
believed in the superiority of "white" Aryan blood, had high
hopes that the accomplished long jumper Lutz Long would secure the gold medal for the "fatherland." The rest, as they say,
is history. Jesse Owens swept to victory in the 100 meters, 200
meters, relay, and long jump, outjumping the German hero by
nineteen "long" centimeters and setting a new world record.
Hitler refused to shake Owens's hand. The Germans had a poor
showing on the track, with only a bronze medal in the relay, again
behind Owens and the US squad. In the field events, they fared
well only in the artillery events, capturing shot, hammer, and
javelin gold to add to Long's silver medal.

tennis balls after using them only once. Karl was able to take advantage of
this windfall by selling the balls.

From 1936 to 1939 Karl participated in weekend and after-school Hitler
Youth activities, those being the only organized youth activities that were
legally sanctioned. One of his interests at that time was the construction
and testing of gliders, and during these aeronautic activities he built craft
with very simple designs that were little more than a seat with wings, a stick
fuselage, and a tail. There was no cockpit or enclosed fuselage to protect the
pilot, but at least the pilot had an unobstructed 360-degree view. Karl never
took these gliders up high, but he got the feel for flight with a series of short
hops close to the runway. Other boys, with less aeronautical skill, provided
Karl with a comic spectacle as they jerked the joystick back too rapidly, nosed
up steeply, stalled, and "stood the gliders on their tails."

When the Summer Olympics was held in Berlin, Karl, who was more interested in middle- and long-distance track events, saw no German successes in
track arena, nor could he get into the main stadium. He did manage to get in
and see some gymnastics and diving, however, with the help of tickets and
passes provided by American visitors. In these sports the Germans fared
somewhat better: Alfred Schwartzmann won three golds and two bronzes
and carried the Germans to the gymnastics team victory, and one of the six

Karl's 1936 painting of a javelin thrower is one of many artifacts that remind us of a controversial and historic Olympics. Original in the University of Colorado Museum archives.

diving medals, a bronze, went to German platform diver Hermann Stork. Karl was quite taken with all the international sports activity and drew a colorful picture of an athlete surrounded by flags and Olympic rings.

He was able to view some of the events on a prototype television that had been installed at the post office. Television was still an extreme novelty in its experimental stages, and it displayed a spiral pattern of dots, or pixels, unlike the rectilinear pattern familiar to later generations of viewers. It was through this new medium that Hitler shamelessly broadcast Nazi propaganda. These were the first television broadcasts ever released on the airwaves, and as Carl Sagan was to point out two generations later, they could be the first such signals from earth to be received by intelligent life elsewhere in the universe.[4]

In 1936, Karl's father was able to afford the fees to enable his son to return to the Friesen Oberrealschule for the last two years of his formal education. When Karl graduated, in 1938, conscription into some form of military or supporting service was all but inevitable. In March, following an ultimatum from Berlin, Nazi forces seized Austria, and installed a Nazi administration. The swastika flew over public buildings. Angry Nazi mobs invaded the Jewish quarter in Vienna, and Jews were forced to scrub the pavements and clean the public latrines. Worse still, they were systematically stripped of all their possessions and jailed or forced to emigrate. An "Office of Jewish Emigration" was established, with none other than Adolf Karl Eichmann (Otto Eichmann's father) in charge.[5]

With Austria in Nazi possession, Czechoslovakia was now surrounded on three sides. Although restrained momentarily by strong protests against military aggression from Britain, France, and Russia, and by Czech mobilization of defensive forces, Hitler vowed that "Czechoslovakia shall be wiped off the map."[6] This he achieved through the Munich Agreement, and through the unprecedented concessions made by France and Britain to hand over the "German" Sudetenland area of Czechoslovakia, in a futile attempt to avoid war at all costs. Prime Minister Neville Chamberlain returned to London from the Munich "sell out" speaking of "peace with honor" and "peace in our time" while, in the face of severe denial and enough political confusion to go around both the French and British governments, Winston Churchill announced, "We have sustained a total, unmitigated defeat." Hitler agreed with Churchill that his British and French adversaries were spineless "worms," all too easily persuaded to capitulate to his demands. Hitler had thus established much of what would become known as the fearsome Third Reich. With perverse logic, after having been given half of Czechoslovakia Hitler lamented that he had achieved only a "partial solution" and that he had been deprived of the opportunity to take the entire country by military force—and so he proceeded to pursue a "total solution." In March 1939 Dr. Emil Hacha, president of Czechoslovakia, visited Berlin and was forced by German ministers to either capitulate to complete surrender or face the immediate bombing of Prague. The next day German troops rolled into the country, and Hitler announced that "Czechoslovakia has ceased to exist."[7]

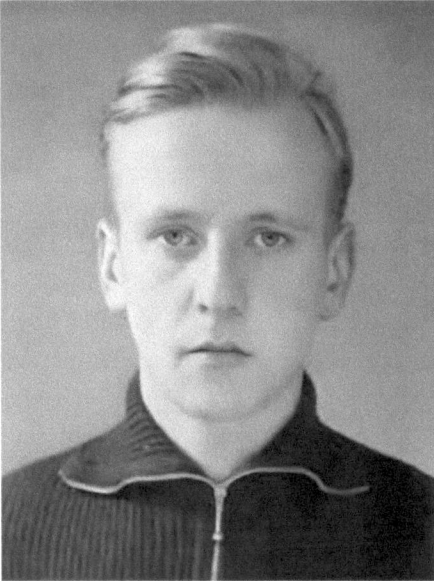

Portrait of a young athlete. Karl was a good long-distance runner and also enjoyed the team sport of rowing. Photos from Karl's own photo album, replicated in the University of Colorado Museum archives.

Such was the situation in the year that Karl left school; it was inevitable that he would be called up to serve the voracious interests of the Third Reich juggernaut. The only way to exercise any measure of choice was to volunteer for a preferred branch of service. Karl tried to pursue his interest in aviation by signing up to serve in air force communications ("Nachrichten") in the antiaircraft division ("Flakregiment"), but he was rejected on the grounds that he was too short. At that time he stood a little less than 1.65 meters tall (5'5"). He was also recruited by the air force ("Luftwaffe"), but after being subjected to a test flight in a Messerschmitt, Karl found he had no aptitude for aerial acrobatics and opted to remain with his feet firmly on the ground.[8]

Karl's father and various cousins had gone into the postal service, so he too joined for a short career as a "man of letters." His apprenticeship with the Civil Service began in May of 1938. As a lean, athletic youth of seventeen, he joined the Postal Service athletic club ("Postsportverein") and began running competitive middle- and long-distance races ranging from 1,500 meters to 10,000 meters. He ran the latter in a respectable time of about thirty-nine minutes, and so he usually placed respectably among the finishers. His small build was well-suited to the longer distances, and running cross-country in winter would later stand him in good stead as a soldier. With youthful energy to spare, Karl also took up competitive rowing and found that it helped build team spirit.

A year later Karl took his annual physical examination and had grown a full fifteen centimeters (six inches) and stood 1.80 meters tall (5'11"). Had this growth spurt taken place twelve months earlier, he might already have been serving in the military communications sphere. Karl's postal training, however, and his participation in this essential service earned him a temporary respite from conscription. But the imminent outbreak of war ensured that he would soon be pressed into compulsory service. It was only a matter of time before his summons would arrive.

Karl, like many a healthy young man, had run headlong into sports and outdoor adventures and was in prime physical condition. What he could not fully appreciate, although the ominous warning signs were all around, was that he was about to run headlong into history, and that meant headlong into battle. The world conflagration that was brewing would change individuals and nations and rudely wrench the destinies of almost everyone involved.

Notes

1. "Die Fahne hoch, die Reihen fest geschlossen" are the opening words of "Horst-Wessel-Lied" (Horst-Wessel song), the nationalist anthem of the Nazi Party from 1930 to 1945. Literally, the words translate to "Raise the flag (high), the ranks tightly closed." Like Hitler's book *Mein Kampf* (chap.1, note 1), the anthem has been banned in Germany and Austria since the war. "Deutschland, Deutschland über alles," the acceptable national anthem, dates to 1922 and does not have objectionable Nazi connotations.

2. Karl, like many paleontologists, had an artistic streak. It was first manifested in three colorful paintings that survive from his teenage years, depicting Berlin's 1936 Olympic stadium festooned with a dozen national flags as well as the Nazi swastika. In one painting he chose to depict the javelin event, won by Germany, with an athlete poised to throw. Ironically, the javelin label would be adopted by the US military in the twenty-first century to name a deadly antitank weapon used, most notably, in the Ukrainian resistance against Russia in 2022 and later. This conflict today is being played out on the same terrain where Karl would be sent in 1939–1940 to fight the Soviet Red Army.

3. A preparator prepares scientific specimens for study. In paleontology this often involves removing delicate fossils from rock-hard matrix. Expert preparators are considered talented craftpersons.

4. In 1974 Carl Sagan famously sent a radio message, named "Arecibo" after the Puerto Rican radio telescope, out into space. Since the Berlin Olympics had been the first widely broadcast television program, Sagan wondered if Hitler would be the first person to be seen by extraterrestrial life (https://www.realclearscience.com/blog /2013/09/will-hitler-be-the-first-person-that-aliens-see.html).

5. The notorious Central Office for Jewish Emigration ("Zentralstelle für jüdische Auswanderung") was established by the Nazis in 1938 in Vienna (and Prague and Amsterdam), for the express purpose of expelling Jews from Nazi-controlled areas.

6. In Berlin on May 28, 1938, Hitler told his generals, "It is my unshakable will that Czechoslovakia shall be wiped off the map." This quote has come to epitomize the so-called Munich Crisis or Munich Agreement, in which Allied forces capitulated to Hitler's takeover of the part of Czechoslovakia known as Sudetenland.

7. Max Domarus, *Hitler: Speeches and Proclamations, 1932–1945: The Chronicle of a Dictatorship* (Wauconda, IL: Bolchazy-Carducci, 1990).

8. The Messerschmitt (loosely translated as "knife smith") was the iconic German fighter plane, probably known to most schoolboys of the war and postwar generations. With the connotation of a cutting weapon, it figured in legendary ace air battles against the British fighter plane called the Spitfire.

CHAPTER THREE

War Clouds

In the spring of 1939, following the takeover of Czechoslovakia, it was obvious that Hitler would next invade Poland. In fact, he had already stated that it was necessary to expand German living space and "attack Poland at the first suitable opportunity."[1] Chamberlain, in London, awoke to the reality of the situation and pledged support for Poland. Battle lines began to be drawn between the Reich and the West, and day by day it appeared that the outbreak of a European war was inevitable. On September 1, 1939, the German army invaded Poland. On September 3, when Germany refused to suspend its aggression against Poland, both Britain and France declared war on Germany. The shock waves rippling across Europe would soon be felt worldwide.

Germany and western Europe, home to Copernicus and Kepler, are situated close to the zero-degree, or "prime," meridian passing through London. Thus, as geographers, historians, and military strategists recognized, the area symbolically represented the Eurocentric center of the solar system and its inner planets. Hitler's war would expand west to the Atlantic and the shores of North America, east across Asia to the Pacific, north to Scandinavia and the Arctic, and south to the Mediterranean and North Africa. Like the impact of a projectile hitting calm waters, or the ripples caused by the first war salvos, shock waves propagated outward in the early war years—until they encountered resistance and rebounded. Simple physics and the laws of entropy would ensure that the shock waves, spreading from the aggressor's fiery source, would weaken—attenuate like the dissipation of energy across the ever-widening rings of planetary orbit with distance from the sun. Consciously or unconsciously, the solar system metaphor wove its way into the strategic language used to represent the physics and gravity of troop movements across eastern Europe. So, after the Axis powers lost eastward momentum, when the Soviet forces rebounded from the outer orbit of Axis influence, the counteroffensive was first named Operation Uranus, which in turn became Operation Saturn as the rebounding energy closed centripetally back toward western Europe.[2] Similar phases of expansion and contraction, advance and retreat, are nothing new in the history of warfare and the clash of civilizations, as Napoleon and others with imperial ambitions had learned from often bitter experience.

For the soon-to-be foot soldier pawn, uneasily sensing momentous upheaval rippling through prewar society, any big-picture analysis or long-term future predictions were little consolation. Philosophy and the wisdom of historical perspective would not likely change short-term outcomes. Resistance to the darker forces would prove difficult and would force crises of conscience that would divide families, communities, and nations. Talk of imminent war had indeed proved prescient; Germany was at now at war, and its Axis allies and enemies were rapidly being defined and delineated, as the calls of fear, conscience, compromise, and expediency would dictate. Events in western Europe's inner orbit were being driven by dark forces: the whims of a self-appointed führer with delusions of grandeur, destiny, and dictatorship that would keep historians supplied with books of commentary and analysis for generations to come. Hitler, whom some have credited with the bizarre belief that the world was hollow, with humankind living on the inner surface of its crust or shell, looking inward to the heavens, had literally turned the world inside out and made heaven the underworld. He had created a hollow universe in which his nation had lost its orientation.

Karl's summons, or *Geftellungsaufforderung*, arrived the following year, dated June 8, 1940, to coincide with the early termination of his civil service training. The summons called him up for compulsory "voluntary labor service" in the *Arbeitsdienst*, an auxiliary organization that helped with the war effort by digging ditches, farming, and helping with various military duties such as supply and camouflage. For the first time Karl wore an armband with a swastika. The weapon of choice was a spade, shoulder-carried like a rifle when marching or called to attention.

Conscripts were subjected to a typical pattern of military training, discipline, and regimentation. They slept in barracks and were roused at dawn to run a mile or two. This exercise was rendered uncomfortable by the lack of available time, or opportunity, to relieve oneself before the jog. By the time the troops arrived at their destination and received the order to fall out, bladders were bursting, and great relief was had in the dash to water the bushes and undergrowth.

Karl was posted to the island of Texel, then in occupied territory in Holland, where his duties involved camouflaging airplanes that flew the first bombing raids on London and painting and maintaining the radio towers. He saw no action at this time and enjoyed the luxury of a brief flirtation with a Dutch girlfriend, skating excursions on the Dutch canals and waterways, and weekend leave in Rotterdam. On such occasions, the young Arbeitsdienst recruits ran into real soldiers and sailors, who were emboldened by a few months of experience in real action. Sometimes fights and scuffles broke out as the youngsters tried to defend themselves against taunts and insults. In such incidents the older and slightly more seasoned troops did not always prevail.

Karl, now nineteen, was bigger and tougher than he had been as a diminutive seventeen-year-old long-distance runner. He began to learn some of the rudiments of self-defense from his friend Helmut, a boxer of some ability. Helmut taught him to spar, and once suffered the indignity of a sharp punch to the nose when Karl caught him with his guard down. Helmut retaliated swiftly with a well-planted knockout punch, leaving Karl dazed, floored, and with renewed respect for Helmut's boxing skills. On another occasion, at a local dance, when a belligerent sailor attempted to pick a fight with Karl, Helmut made a timely appearance on the scene. The dispute was settled quickly when Helmut floored the sailor with a single decisive punch. Youthful pugilism was just one species of belligerence sweeping Europe.

At this time Karl also met his friend Christoph, who was a grandson of the famous German actor Paul Wegener.[3] Coming from a family with more than its share of artistic genes, Christoph, unlike Helmut, was ill-suited to military parades and aggressive pursuits. His father was a university professor and was later to write letters of support for Karl when he ran into bureaucratic problems after the war. Karl was close to the family and, having his own share of empathy and creative sensibility, kept an eye on Christoph so that his dreamy artistic temperament didn't get him into trouble. Karl and his companions made sure Christoph made his bed properly, in time for inspection, and did not fall foul of the authorities for failing to conform to disciplinary demands.

Ill-suited to frontline action, Christoph was later to be an early casualty on the Eastern Front.

After five months in the Arbeitsdienst, Karl, Christoph and other comrades were called to Berlin for a three-month season of basic training. Here they learned to handle weapons and take care of horses and military equipment. During air raids they would have to take horses to the woods and try and keep them under control as they shied and reared in reaction to bombardments around the city. Failure to keep the horses under control would result in punishment in the form of extra duty. Far from being good at this type of activity, Christoph was barely capable of turning out for a parade in proper attire. On one occasion, when the troops turned out to see Hitler, Mussolini, and the Japanese leader Tojo on parade, Christoph managed to polish only one boot. This minor offense against the laws of symmetry by a dreamy artist threw the master sergeant into an asymmetric fury.

By now the war in France was over, and everyone knew, despite the previous Stalin-Hitler accord, that war on the Russian Front was inevitable. There was talk about how Russia would be much harder to conquer and impossible to successfully occupy. Although it was a banned book, Karl and some of his comrades had read *Der Untergang des Abend Landes* (*The downfall of the western world*), which discussed the folly of Napoleon and other would-be conquerors of Russia and Asia. Such politics and strategy were discussed by Karl and his fellow foot soldiers when, on weekend passes, they took their growling stomachs to the local *rathskellar* to devour all three meals on the menu.

Notes

1. Wilhelm Keitel testified at the Nuremberg trials that Hitler had stated on May 23, 1939, that he had decided "to attack Poland at the first suitable opportunity."

2. Sources on these Soviet counteroffensives are too numerous to mention. A representative title, by David Glantz and Jonathan House, is *When Titans Clashed: How the Red Army Stopped Hitler* (Lawrence: University Press of Kansas, 1975).

3. Paul Wegener (1874–1948) made his mark not only as an actor but also as a writer and a director of German expressionist films.

PART 2

Prisoner of Fate

CHAPTER FOUR

A Journey of a Thousand Miles

Those who cannot remember the past are condemned to repeat it.
George Santayana

Winter came. Bad weather, cold weather.
Karl Hirsch

Karl's history lesson began in May 1941 when he was assigned to the Sixth Army and posted to the front as part of Operation Barbarossa, which notoriously broke the German-Soviet nonaggression pact. Germans involved in Hitler's invasion, if they were not early casualties, would have plenty of time to learn and ruminate on the so-called lessons of history. They would also, as Karl did, refer to the Soviets as Russians.

In June, Karl traveled by train from Berlin to the Moldovan region of northeastern Romania, around Botoşani, a distance of six hundred miles. The southeasterly route from Berlin led through what had been southern Poland and its borders with Czechoslovakia, and then across Hungary. The Botoşani region has a long history and was familiar with invasion. Indeed, it is thought to have derived its name from the Tatar chief Batu Khan, a grandson of Genghis Khan who had invaded the region in the thirteenth century. A few dozen kilometers from Botoşani are Europe's Carpathian Mountains behind and to the west, and the vast Asian steppe in front. Looming over the whole region are the ghosts of history, invasion, and human ambition.

The invasions of the Soviet Union by Hitler in June 1941 and Bonaparte in June 1812 both turned into tactical disasters when mild summer weather turned to winter, trapping both armies in Russia, where they were overwhelmed by tough, zealous soldiers defending their own soil.

It is symbolic of the historical, political, and ethnic divisions between Teutons and Slavs that the Germans referred to their Soviet adversaries as "Russians," thereby geographically underestimating the ultimate power and reach that the Red Army (the Workers' and Peasants' Red Army) would exert in defending the vast Soviet territory. For the Soviets, their resistance to Germany would be their "Great Patriotic War." Among the soldiers, a million women took up combat roles. The term *red* symbolized bloodshed in the fight against capitalist oppression. The ordinary "Russian" rifleman would be nicknamed "Ivan," just as the German foot soldier would be "Fritz" and the British infantryman "Tommy."

The Prut River, marking the border between Romania and the Soviet Union, also marks a larger geographic frontier between Europe and central Asia. Karl, there as part of a large troop buildup with continental-scale ambitions, would begin a long march, indeed an odyssey, that would take him more than a thousand miles, across Ukraine and all the way to Stalingrad. He would participate in what has been called the greatest war in recorded history (Lucas, 1979). Operation Barbarossa, the code name for Hitler's invasion of the Soviet Union, would develop a north-south battlefront of more than one thousand miles along the western border of the Soviet Union. Barbarossa, meaning "red beard," was a nickname given to Frederick I (1123–1190), who was both a German king and a Holy Roman emperor (1155–1190). He died during the Third Crusade while trying to cross the Saleph River in Turkey.

The southern advance began on June 22, 1941, with the crossing of the Prut River. The Soviets were prepared for the onslaught and put up strong resistance as the first wave of German troops came across the river in speedboats. Many in the first wave were shot, shelled, or drowned under the weight of arms and ammunition belts, and it was not until the second wave landed that a beachhead was established. It was in this second wave that Karl crossed the Prut, the first of many rivers, and saw his first action.

Decades later Karl would reluctantly commit his traumatic memories of these events to tape. What he saw that day was a "terrible sight . . . something I will never forget."[1] The sudden attack took a heavy toll on both German and Russian soldiers. "My platoon was in the second wave," he recalled. The first wave had been worse, with "boats blowing up and bodies flying everywhere. Few from the first wave made it across. They never trained us for that." Improvising desperately, those in his platoon decided they "should untie our boots and unfasten our packs" to prevent drowning if their boat capsized. Even though Karl's boat made it across, "the boat behind . . . was hit and the fellows went down."

"Fellows" seems to be a word of affection and common experience that rises to the surface of war stories and tales of institutional comradeship, whether recalling fellow comrades and brothers in arms, fellow prisoners, fellow athletes, fellow students, or even fellow academics. Although Karl and his infantry comrades were cannon fodder for the *blitzkrieg*, the concentration of air and fire power, these German shock-and-awe tactics "forced the Russian army to retreat," relieving some of the immediate pressure on men like Karl who had survived the first onslaughts. But the gates of hell had been opened. "That day was," Karl recalled, "the first I saw a dead soldier." He "never got used to seeing the dead. . . . I just got numb to it."

Obedient to orders, they had established a beachhead on the eastern bank of the Prut. Karl and his friend Christoph served as messengers and ordinance men, running messages from the battalion commanders to various platoons. On one of their first forays through the woods on the east side of the Prut, they came face to face with the enemy, a lone soldier armed with his rifle. Instinctively, knowing that Christoph would be slow to react, Karl shot from the hip, hitting his foe in the gut. With an utterance foreign to the German ear, the man went down. To the Axis powers he was an unknown soldier, an unknown casualty, perhaps even named Ivan. Early in the southern campaign, Karl had fired his first fatal shot in pursuit of the expansionist objectives of the Third Reich. It had been one thing to see killing for the first time, another to kill, even in self-defense, even "when you had people who depended on you to survive, and . . . when . . . you depended on them as well."

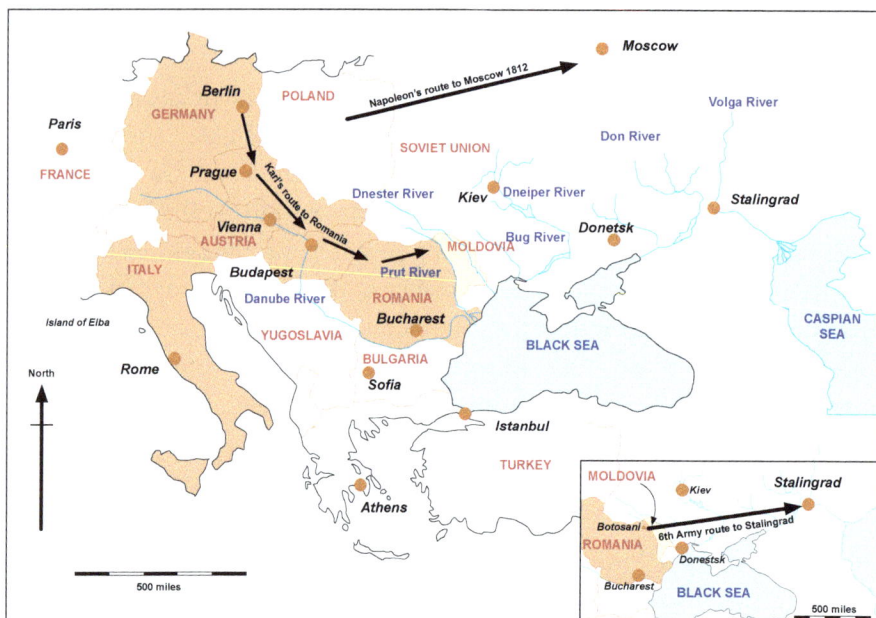

The big-picture map of Karl's quick train journey from Berlin to eastern Romania, and the territory that lay ahead (inset) as the German Sixth Army marched to Stalingrad.

Karl's infantryman education was swift and brutal: "Suddenly you find you have 'actually' killed someone face to face . . . a really bad feeling . . . as if you are some kind of animal." Seeing his enemy "all torn open" and moaning a little left Karl, as he would put it later, with "knees . . . a little soft."

In wartime, maps and territory change with maneuverable speed, keeping geographers, cartographers, historians, and higher-ranking officers busy with pens and little flags. The strange masculine fascination with war ensured that considerable ink, much of it red, would be dedicated to battlefront maps. Front lines are like ripples, and troop movement arrows mirror projectile trajectories, the proverbial "slings and arrows" of outrageous fortune. Successive maps of advance and retreat resemble the patterns made by the swash of waves on a beach at the turn of the tide, with larger and smaller arrows showing the various more-or-less powerful currents. Indeed, the wave and tidal metaphors are age-old in military history. With the advance of a rising tide, waves bulge and spread forth to reconfigure the crenulated frontier

July 14, 1941. Three weeks into their offensive, German foot soldiers made rapid progress eastward across Moldova and Bessarabia. Marching helmet-free, they faced little resistance. They would reach the Bug River by early July and the Dnieper River between Kremenchuk and Dnipropetrovsk by early September.

of land and sea, or of aggressor and defender forces, until the aggressor's impulses weaken, the tide retreats, and the defender literally reclaims lost ground. Here and there, flanking backwash eddies add to the picture of dynamic chaos.

Advancing east of the Prut into the Soviet Socialist Republic of Moldova, and then across the larger Dniester River the Sixth Army made rapid headway through western Ukraine, as the Soviets retreated. The advance to the Dniester River was effectively the wave that pushed the boundary of Romania fifty to one hundred miles east of the Prut, creating an area with the Middle Eastern or Asian-sounding name of Bessarabia, echoing a complex history of territorial wars between Russia, Turkey, and European powers dating back to the time of Napoleon. Soon Karl was among a tide of troops, in the Balta region east of the Dniester, halfway to the Bug River, a large new slice of river-bounded territory that would become known by August 19, only two months after the initial advance, as the Romanian Transnistria

The Prut, Dniester, Bug, and Dnieper Rivers, as well as the Don and Volga further to the east, would play, as they had in the past, pivotal roles in defining the regional geography and military history. As simple and fast as the advance from Romania to Balta might appear, the short-lived history of Transnistria, which was reconquered by the Soviets in January 1944 after an occupation of only twenty-nine months, was anything but simple. Hitler, in his bid for Romanian support, had made it easy for the German army to capture Bessarabia, some of it formerly Romanian territory, by also advancing down the east bank of the Bug River and forcing the Soviets on to Odessa on the shores of the Black Sea. Odessa was a city the Soviets had defended fiercely at the expense of many Romanian lives. In retaliation for the Soviet resistance— including a literal and very effective Soviet underground in the Odessa catacombs—Romanian leader Ion Antonescu massacred thousands of civilians and deported many Odessa Jews to concentration camps. More ghosts would loom over the region for generations to come.

Governate, or simply Transnistria, Nistria being the Romani name for the Dniester River.

As "hard fighting" continued, it was only a few days after Karl had shot his first enemy soldier, perhaps saving Christoph's life as well as his own, that he would lose his friend to a mortar. Karl quickly learned how the shells would "whistle on the way down," which was the signal to "dive for cover." Sadly, Christoph was not a "good soldier," despite dying for someone else's cause. Karl described that if you had "to think about it, it's usually too late. . . . So, the mortar got him and he was all covered in splinters. I patched him up as best I could and took him to the field hospital where he later died." Not only is losing "over twenty percent of our men" to "hard fighting" very hard to bear, but "it is very hard to explain what it was like if you are talking to a person who has never been in war. The experience hardens you. You are never quite the same person again."

Notwithstanding the initial resistance faced at the Prut River frontier, while Karl and his comrades marched often, with comparative ease, east across the wide-open steppe, the Romani-German tide had crashed up against the rocky concrete defenses of urban Odessa. This was of no immediate concern to Karl, who was well to the north. The few pictures he obtained from these early days give the deceptive sense of young soldiers embarked on a relatively carefree adventure. But while unresisted advances were moments of calm between storms, reality spoke of what was already a hellish adventure, with the blitzkrieg's flames burning into flesh and psyche. Karl would sum up what commentators describe as the "futility of war" with the foot soldier's perennial wisdom: "I never felt like I had to avenge my friend's death, nor . . . feel a sense of anger towards the soldiers we were fighting. . . . After all, this is war we are talking about. Being in that situation of having to march mile after mile and follow orders, of having to kill in order to survive, of having your friends die in front of your eyes . . . there was no time to mourn the death of a friend or feel sorry for yourself."

In the early days of the campaign, Karl and his comrades passed through German-speaking areas, where the advancing troops were often welcomed with flowers and treated as liberators. Flowers were thrown and buckets of water and milk were left outside houses to refresh the thirsty foot soldiers slogging east. Some simple, barefoot peasants "stroked the black crosses of German lorries, convincing themselves that the men whose army carried this holy symbol . . . were their allies in Christ."[2] There was little partisan activity in this region in these early days. Not until the occupation troops arrived with their oppressive methods were resentment and strong partisan resistance inevitable. With the rapid German gains, many frontline Russian infantry captives, who had surrendered with little or no resistance, were swept along with the advancing German tide. Generally, they were treated humanely and were expected to do little more than march along and help carry supplies. If they then reached their home villages or provinces, they were often released, though to do so was technically an act of treason on the part of their captors. As Karl would later tell,

> When we were advancing we took a lot of prisoners. Sometimes . . . to carry the ammunition or keep watch while you took a nap. Of course, you picked the ones you thought you could trust. Most . . . wanted to stay with us because we treated them fairly. . . . They would help us forage for food. . . . On occasion . . . a fellow would say "I live just over there," and we would let him go. We were all just *armes schwein*, poor pigs together. Hitler, he did not treat them well, so they were afraid of the people in the rear. You see the fellows in the front, they're in the same boat you are in. They got

drafted; you got drafted. They had a home and a family, perhaps a little boy or girl . . . just like we did. When it was time to fight you knew he has to kill you, or you will kill him. . . . It's a strange relationship, but you knew where you stood with that fellow. The people in the rear, the administration . . . some of them had never seen the front. . . . You were afraid of those type of people.

Once across the Dniester and Bug Rivers, the Sixth Army drove east toward the city of Krivoj Rog, which was occupied by mid-August, and then marched for Nikopol on the Dnieper River, where Hitler had his eye on the strategically important manganese deposits. Nazi killing and imprisonment of Jews would begin in October, after Karl had skirted northeastward to meet the Dnieper River further upstream.

Two to three months into Karl's foot soldier experience, the war was intensifying. The few remaining low-resolution photographs of Karl from this period show him becoming a seasoned soldier. One snapshot seems to capture his Ukrainian experience: he sits in a large bed of straw, weather-beaten and unsmiling, a hunk of bread in one hand and a canteen, probably containing only water, in the other. Summer is ending, the steppe is stubble, straw mounds stretch into the flatland distance. A German boy, not yet 21, was in foreign territory somewhere between Kremenchuk and Dnipropetrovsk.[3]

Years later Karl would say of the landscape, "It looked a lot like Kansas." Perhaps, as in the legendary *Wizard of Oz* story where tornadoes had disrupted bucolic peace, Karl, always one for wide open places and spaces, would feel a certain nostalgia for moments of communion with expansiveness and calm between storms. Both the Kansas prairie and the Ukrainian steppe had subtle heartland powers to stir the heart of the nomad soldier. The German soldier-poet Erich Swinger, who we shall meet later, would express such bittersweet sentiments.

So much for the grounding a simple recruit might feel in the agricultural steppe. As Karl moved east, "We left the agricultural area and entered an industrial area near the town of Dnipropetrovsk. By this time, we had walked about a thousand miles. We still had Russians on the run and were advancing quickly, covering a lot of ground. . . . Sometimes we would even march at night, clearing out Russians as we went."

Marching is the bane of the foot soldier, and following a Panzer division that could advance thirty miles or more a day called for determined foot-slogging just to keep up: some claimed daily marches of seventy kilometers,

Bread, water, and straw near the Dnieper River between Kremenchuk and Dnipropetrovosk.

almost forty-five miles. After a hard day's march, the men would collapse half-exhausted on the roadside, only to be roused again after a short respite. As they tried to sleep, they would first hear the distant call of commanders rousing their units to action further up the line. Desperate to rest and cling to slumber, Karl and the men of his unit would close their eyes more tightly, willing themselves to shut out military obligation for a few more minutes. But the effort was futile and the calls to action became closer and louder, amplifying their way inexorably down the line. On the move again, the infantry would sometimes march blindly though fields of sunflowers that could easily hide a whole battalion. Foot soldiers would fan out to guard the flanks.

Compared to a river crossing of less than a mile, which might take up to a week, each wave of advance on dry land between Ukrainian rivers was swift, up to thirty miles a day. The large rivers obstructing the line of advance provided opportunities to wash off the grime of war. Photos of Sixth Army soldiers on the banks of the Dnieper in early September show a platoon-sized group naked and ready for a dip. The Dnieper is one of Europe's largest rivers, with a drainage of more than a half million square kilometers. It occupies a large part of present-day Ukraine and much of Belarus, with headwaters extending north to the borders of Poland and Lithuania and almost to Moscow. The Dnieper gives its name to Dnipropetrovsk, a town remembered by Karl and fellow soldiers, on both sides, and a place where, Karl would recall quite simply, "Our advancement stopped." Natural landscapes and riverscapes, reclassified by politics and war as territories, frontiers, and boundaries, would again remind armies, made up of mere mortals, of the challenges nature poses to vain human ambition. As if to underscore this reality, Karl would later recall, "Then the winter came. Bad weather, cold weather."

German efficiency was not to be stopped on the western bank of the historic and prehistoric Dnieper. It had only been delayed. Photos show some of Karl's comrades standing naked on the river's shores, washing off grime, their belt and high-thigh tan lines suggesting they had found time to sun themselves on their march across the steppe; there was perhaps, as yet, little thought of the frigid winter to come. Their orders were to press on. Other photos show troops, now in full field gear, ready to board the rubber boats that would ferry them to the eastern side of the river. The engineers would follow with the pontoons necessary to bring across the Panzer division tanks.

The Germans would cross the Dnieper River by August 31, just south of Kremenchuk almost two hundred miles downstream from Kiev, where the river was twelve hundred yards (three-quarters of a mile) wide. Tactically the crossing was a triumph of German organization and stealth, backed up with air support that ensured they knew where the enemy defenses were on the east bank. The Germans used two types of boat: a smaller one capable of carrying four to six men and large ones that could carry ten to sixteen men. As the boats reached the east bank, the crew did not stop but had the men immediately jump out—to avoid making a static target. They quickly turned the boats and headed back for the next load, crossing the river again and again. While much is not known of the hour-by-hour day-by-day activity, we know from Karl's meager photo collection that he stood on the banks

It is unlikely that Karl knew that the Dnieper River was the product of the melting of the last Scandinavian Ice Sheet, which had spread south from the Arctic, reaching present-day Belarus and creating a bleak, wet outwash plain during the Last Glacial Maximum, when the vast area was home to both the wooly mammoth and the rhinoceros. Nevertheless, the river's glacial history would revisit its chilly legacy on German and Russian troops alike. The river's meltwaters and decaying ice sheet had created extensive boglands, the vast forty-thousand-square-mile Pripet Marshes. This area was not just the source of much of the river's water today. As a subject of study for geologists, who know the river as the prehistoric link between the Black Sea and the last great ice sheet, just as the Volga plays a similar role as the link between the ice sheets and the Caspian Sea, these marshlands were also a daunting obstacle for the German armies in Operation Barbarossa. In their northern rush towards Smolensk and Moscow and their southern push for Kiev they avoided the treacherous marshlands, ultimately leaving them as a refuge for partisan resistance.

These mighty rivers had also played famous if controversial historic roles, allowing the Scandinavian Vikings access to what would become Russian territory. To be more precise, the Rus' people, or the Viking Rus, are in many ways synonymous with what western Europeans like to call the Vikings of the eighth to eleventh centuries, and what the world knows as the Russian people and the people of Belarus. While these early Viking Russians or Russian Vikings were not known for leaving written records, coin hoards, and other artifacts show that they traded as far as the Black and Caspian Seas, and it was Persian geographers of the ninth century who first described these traders bringing furs and other goods from northern Europe as far as Baghdad. Karl and his German fellows were up against descendants of Viking stock, who like the mighty steppe rivers had flowed out of the frozen north. But this time they had not come with furs for trade or with offerings of insulation against the glacial chill.

German troops preparing to cross the Dnieper River in the first days of
September 1941.

of the Dnieper no later than September 2. He was now ten weeks into the
campaign, three hundred miles and four rivers east of the Romanian frontier.
Apart from being a capital offense, retreat, like desertion, was impossible
unless one was wounded.

Karl would no longer be marching through fields of sunflowers, welcomed
by unarmed Ukrainian citizens, or sunning himself in golden haystacks
while willing prisoners helped with the heavy lifting. More often it was time
to dig in, to go to ground in foxholes. As the mighty Dnieper rolled on, as
it had for the last twenty-five thousand years, the random futility of war
was manifested in daily incidents across the front from Kiev to the Black
Sea. Karl had been playing cards in a foxhole with two companions. For no
conscious reason he left them to return to his own bolt-hole in the steppe.
Five minutes later a Soviet mortar shell struck his companions' den. He went
to check on them and found them both dead. On another occasion he left
his foxhole with a set of orders to deliver and walked through cover down a
wooded creek. But he was seen and shot at. Scared but unscathed he ran, as
he recalled, "like hell."

Strangely, a story is told of Hitler

Karl's march from Romania to Stalingrad ended just west of the city, with his first wound, which gave him his lucky *heimatschuss*, or "home shot." By then he had walked more than a thousand miles, crossed six major rivers, and seen more than a year of bloody combat. After spending the 1941–1942 winter near Donetsk, he could not know that more than a year later he would find himself there again.

> asleep in . . . the German trenches when he dreamed that he was about to be engulfed in an upheaval of earth and mud. He broke out of this nightmare [and] . . . feeling suffocated and fighting for breath, he stumbled out of the dug-out for air. He had hardly got clear when an enemy shell hit the post and killed all his companions. He himself looked upon this dream as an act of Providence intervening to save him for a greater destiny and from that moment [succumbed to] the conviction that he was under the special protection of fate.[4]

Karl had little in common with Hitler, beyond being a German involved in war, but fate perhaps dispenses its dreams and intuitions impartially. Karl would survive to live out a meaningful destiny.

Karl would dig in at Artemovsk (Bakhmut) near Donetsk, east of the Dnieper. It was a challenge just to keep from freezing to death. German troops, as Karl recalled, were "not that well equipped," so "when we had prisoners, we took their winter coats . . . their hats also . . . hats with ear-muffs." In retrospect, these forced and dangerous wardrobe exchanges were an ironic "dressing up" in the enemy's clothing, even a dead man's clothing, not for subterfuge but for survival. It was a matter of necessity, when, as Karl remembers,

> My boots just fell apart. . . . In the village there was one Russian soldier . . . [with] . . . felt boots that seemed to be my size. So, he had to take them off, and I gave him mine . . . [which] . . . were falling apart. I had not had my boots off for four weeks. He said "They are too small," and I said "Okay, cut a hole for your toes." That's the way it was. Some people accidentally got shot because of what they were wearing. You had to watch but that was the only way to survive.

Clothing, food, survival. "About twenty miles behind the lines there was an arsenal. They pulled us back . . . for a couple of weeks rest," where Karl and comrades ran into a supply sergeant "we didn't much care for." The ser-geant "wound up dead"—no details given! "So . . . we went in and took the winter clothes." The front lines and access to resources changed as often as the clothing. Troops would not "advance at the same time in a long front. . . . Part of the line would get bogged down and have to fight [a] way through." Then "Russians would . . . get behind our front and cut off supplies." On one occasion, "our mess wagon got lost" and the Russians found it and as "they couldn't cook as well . . . took our cooks along . . . and had a good meal for a few weeks," until "we sent reconnaissance troops . . . and got them back."

It was not uncommon for regular troops to steal supplies from the Schul-tzstaffel—better known as the SS, whose name, ironically, means *protective echelon.* In this regard, as unwitting suppliers to the regular troops they lived up to their name, even though they were reluctant to share equally. War was hell. There was an occasion when Karl lay so close to a wounded comrade that he felt the man's warm blood seep through his clothing and freeze onto Karl's skin. What did it mean to be warm, and what price did the soldier pay?

Karl spent many winter days and weeks proving the stereotype of the sol-dier's life consisting of long periods of boredom punctuated by moments of terror. He and his comrades were hunkered down in wooden-roofed bunkers beside potbellied stove. Rodents plagued the battlefields, feasting indiscrimi-nately on the dead human and horse flesh and invading the bunkers. To re-lieve boredom, the men waged losing rodent extermination games, holding

Infamously, this "greatest war" was won in part by Mother Nature, through factors other than the advance and retreat of troop tides and currents. Mud played a huge role in spring and autumn, sand and dust attacked machinery in summer, and extreme cold played havoc in winter. Temperatures of −40°F made fingers, cheeks, and foreheads stick to metal, triggers, and goggles, tearing away chinks of flesh, retarding the healing of even minor wounds. Rifle firing pins broke like brittle sticks of chalk. Soldiers would fire their weapons just to warm the gun metal and their hands. Freezing troops rarely had a hot meal, as cooked food quickly froze solid between mess halls and any but the nearest trenches or foxholes. Heating fuel in the form of cakes of paraffin wax was often in short supply. Karl was perhaps lucky not to have wintered further north in even more frigid temperatures, where Germans would sometimes look out into blinding blizzards only to see white-clad Siberian troops ski into close range, open fire with machine pistols, and disappear like white ghosts.

a well-chosen log vertically and letting it drop like a silent pile driver on the unsuspecting target.

Years later, as a survivor of many such ugly experiences, Karl would be asked the facile question "What was it like" to go through such horrors and hardship? He would share the details sparingly in a matter-of-fact way but would often lament, "I don't really want to talk about it. It brings it all back, and I don't sleep too well." There were the few bizarre and ridiculous tales, such as when his comrades were requisitioning supplies from the local peasants and an old man gave them plaster of paris instead of flour. They only noticed when they mixed it and it set solid. They sought out the rascal but found him long gone.

In 1942, "After the winter we advanced again," as the German troops advanced toward the Don. Pressing on via the towns of Voroshilovgrad and Millerovo, the Russian defense became more desperate, at times throwing cavalry and women against the enemy. On one occasion, Karl and compatriots were attacked by a saber-wielding cavalry unit as they advanced down a large gully. The fight was one-sided, as the Germans defended with machine guns and pistols, quickly routing the offensive, downing both horses and men. Only one German suffered a saber wound, to his shoulder. Germans were both mystified and depressed by the extraordinary tenacity of Soviet troops advancing wave after wave as cannon fodder, and apparently oblivious to their safely and undeterred by the huge losses they sustained. How could anyone defeat such a determined, stubborn, and patriotic enemy? What way of life or ideals were they fighting for? The Soviets might well have asked the Germans the same question.

The Don, another geological remnant of glacial history, was the last major river en route to Stalingrad. It was the last river the Germans crossed in their fourteen-month drive to this historic destination that would soon be their undoing. In another interlude of geographic recollection, Karl would remember, "The area was open plains so you could see for miles. We came to the Don River . . . [which] . . . is like the Mississippi. It has white banks and white sand, almost like a beach." But like the Dnieper it was an obstacle to German ambition, against which the Axis forces hurled and hurtled their blitzkrieg might, demanding the sacrifice of the luckless foot soldier.

Still east of the river, with "anti-aircraft guns, German 88s" behind them, the troops saw "about 400 Russian tanks come over the hills." Firing over the heads of the foot soldiers, the 88s, as Karl recalls bluntly, "shot all of them down." But still there were "Russian troops that had pushed through the armored troops." With landscape and riverscape ever in play, the Don would prove no easier to cross than any of the other large rivers, and as history would later judge, the crossing would prove a largely futile exercise.

Attempting to cross in their rubber boats, the first wave of units in Karl's sector was decimated to the point of complete extermination. Further to the north, however, the German units had been more successful and had established beachheads on the eastern banks. From there they fanned out, forcing the Russians to retreat and reducing the east bank resistance. Again Karl was lucky, as he had been once before, to be in the second wave, avoiding the terrible fate of the first wave. Nevertheless, the resistance was significant, and Karl's unit was pinned down on the sandy beach all day and could not move out and find safer cover until nightfall.

Over the next few days, the Germans gathered strength on the rolling steppe and worked their way east from the westward-flowing Don drainages into the eastward-flowing Volga system. With little strong resistance they advanced on the final approach to Spartakovka and the northern fringes of Stalingrad. The history books tell us that, several weeks earlier, Field Marshall Kleist had crossed the Don to the south, and that had he not been changing tactics constantly, owing to Hitler's muddled directives, "the Fourth Army could have taken Stalingrad without a fight at the end of July." But in battle, things don't always work out as planned. The Fourth Army was diverted from the south to help the Sixth Army cross the Don further to the north. Valuable time was lost, allowing the Soviets to reinforce and establish the Stalingrad Front, albeit by less than efficient means. Fighting had been particularly intense in the late July and early August summer offensive, with a huge tank battle staged near Kalach, less than fifty miles west of Stalingrad. Although the Germans had five hundred tanks and superior weaponry and technical skills, they had run ahead of their supplies and were forced to postpone their offensive, while the Soviets, still at a severe disadvantage, were nevertheless able to field four hundred tanks and harass the German forces with flanking movements that slowed down their advance. The battle, involving two hundred seventy thousand men on the German side and one hundred sixty thousand Soviets, was not decisive enough to prevent the German advance to Stalingrad. Nevertheless, there were huge losses on both sides, and the reported listlessness of Italian and Hungarian troops undermined German efforts. Despite capturing some fifty thousand Soviet troops, and having superior air power, the Germans had been tested. They would later find that the Soviets had secured a bridgehead on the Don that would eventually be the Achilles' heel that would undo the German Sixth Army.

World War II historians regard Stalingrad as a pivotal turning point in the war. Karl would also find it a pivotal turning point in his odyssey.

Notes

1. Unless otherwise stated, all quotations in this chapter that are obviously in Karl Hirsch's voice come from his brief biography *Walking in Your Moccasins* (WIYM), from the Hirsch Audio Archive (HAA), or from notes made by the authors when interviewing him directly.

2. J. Lucas, *War on the Eastern Front: The German Soldier in Russia 1941–1945* (London: Greenhill Books, 1991), 9.

3. All black-and-white photographs in this chapter come from Karl Hirsch's own scrapbook and are shared with us courtesy of the Steinhauer family.

4. L. van der Post, *Jung and the Story of Our Time* (London: Chatto & Windus, 1975), 22.

CHAPTER FIVE

Retreat, Recuperation, and Redeployment

On the morning of August 24, a few days after the Don crossing, Karl and his platoon were dug into their foxholes on the perimeter of a village when Russian T-34 tanks began to advance. A foxhole is no place to be in such a situation, since with a little practice a tank driver can turn his track on the top of a foxhole and fill the foxhole with earth. Any soldier caught inside will instinctively duck, and so be buried with little hope of digging out. The alternative, to jump out and run for cover, is equally dangerous, presenting the enemy with an easy target. In short, you are in a pickle and must hope for distractions and the chance to take evasive action. Karl would later tell the story of this day as follows: "We pushed through with the tanks to Stalingrad . . . covering the flank . . . I could see over the hill into the valley where . . . Russian troops were coming our way. We never really knew how fast they were advancing. . . . Our platoons had about thirty people . . . divided into four smaller units . . . dug in in front of the village. There were seven of us and I was the leader. We had taken precautions in case they attacked."

As the Russian tanks advanced, the German Stukas retaliated by diving on the tanks to provide a distraction. Since the Stukas were not armed with bombs, these maneuvers were merely rather elaborate scare tactics, at best designed to harry the infantry. After a few passes the Russian counteroffensive continued. Karl and his men prepared to retreat. Even though the enemy planes were not carrying bombs, there was half a chance that the Stukas' swoops would make the Russian tank operators flinch or duck, at least momentarily. Knowing this, and with the tanks now only a hundred meters away, Karl's thoughts flew:

We had to get out of the foxhole because if they saw you they would run over the foxholes and turn and then you are buried. I know. I went through it . . . the hole caves in and you're done. So, I yelled to my fellows, "The next time the bombers come, go into the village" . . . [where] . . . there were some houses and gardens and little trees . . . more cover. The dive bombers came back, and this time got pretty close to the tanks, so we took off. . . . I was good at sports so I was the fastest. Another fellow also good at sports, was not quite so fast. He was a few feet behind me. The tank fired a shot right between us. All of the splinters, the little shrapnel got caught in our packs, ammunition vest and ourselves. I got one splinter in my elbow and another where the buttocks join the leg. I could still run . . . right into the village. The other fellow caught a whole bunch in his arm, leg and all along one side. We both ran into the village and . . . looked each other over. . . . He couldn't sit down right . . . not functioning 100 percent . . . we decided . . . it would be best if we kept running. The other fellows didn't make it. . . . We ran between the houses. . . . Russian tanks in there and there was a big mix up. . . . We went . . . out into the open . . . [and] found one of those washes like you see in Wyoming and followed it uphill. On top . . . were some German officers. They . . . saw that we were wounded and let us through. This fellow and I had to cross through a minefield.

Karl and his friend had "seen mines in dry country, tank mines. . . . They get covered up with dirt . . . and the sun hits it all day long and you know it's there by the cracks. We must have made our miles that day, maybe twenty." Adrenaline-driven, they had run unmeasured miles across the empty steppe, at times slowing to negotiate the telltale minefield cracks and signs of ground disturbance. The two men, one a former 10,000-meter runner, made it to the westward-flowing Don drainages without encountering other units on either side. There they found their way back to the relative safety of a field hospital behind German lines. Long-distance endurance had helped save the day. The two runners "fell asleep right away," exhausted, while the doctors dealt with the more severely wounded. When Karl tried to stand, he found it difficult as his legs ached and throbbed. Even though the outward sign of the wound was slight, and the doctors at first suggested it was nothing and that he could return to the front, he knew otherwise.

He had received what was known as his *heimatschuss*—the "home shot"— the salvation battle wound that one hoped would, like this one, not be too severe. "As soon as you got wounded you got a ticket home. It was the only way. I had started in Romania . . . then all the way to Stalingrad. There was no chance to go on further," as history would soon prove.

Now that the adrenaline had worn off, his leg did not work properly. Karl felt the pain of the shrapnel and knew that he could not function in action in his hobbled condition. "In a way I was happy" but when it was "our turn to go in I was unable to walk." They later found "a tiny little splinter ... right in front of the sciatic nerve." But this *heimatschuss* was to the battle-weary and not-too-severely-wounded soldier like a miracle. "They sent me back to the next hospital, and the next, until I reached Poland." Karl would later write in terse notes for a talk to American high school kids, "Stalingrad . . . got wounded—made 20 miles to field hospital. Hospital was like heaven—nurses like angels."

On that same day of August 24, 1942, the Sixteenth Panzer Division launched an attack on Spartakovka, the northern industrial suburb of Stalingrad, and the next day (August 25) the Regional Committee of the Communist Party in Stalingrad proclaimed that Stalingrad was under a state of siege. During the next five months the Germans occupied more than 90 percent of the city, but the Russians never completely relinquished their foothold on the west bank of the Volga and fought back with heroic tenacity, eventually using the winter, counteroffensives, and Hitler's military miscalculations to surround the Sixth Army, and starve it to death, in what became known as Operation Uranus, symbolic perhaps of Hitler's ignorance of the centripetal forces impinging on the outer spheres of his overextended reach and unrealistic worldview. Hitler refused to believe that Stalingrad would be lost, and, as he had done throughout the war, never let his generals act independently and make the best military decisions. He sacrificed huge numbers of men rather than let them retreat or surrender in the face of hopeless odds. Conditions in and around the city were horrific. The carnage of human and horse flesh was appalling, and the amount of it was so huge that there was a population explosion of rodents. Rats became so bold that they gnawed on emaciated, starving soldiers who barely had the strength to fight them off. When men had picked clean the bones of the horses, they were reduced to eating rodents, human flesh, or nothing at all. The failure at Stalingrad turned the tide of the war against Hitler. By the time the Soviets retook Stalingrad, not a single building remained intact; the whole area was a wasteland of rubble, bones, and frozen bodies; and more than three million men had perished, almost one-tenth of the entire "greatest war" casualty toll.

A "lucky" wound in August had saved Karl from almost certain death in the abominable Stalingrad winter of 1942–1943, in which a quarter million Sixth Army troops were cut off and starved into submission. But although he had perhaps already used up three or four of his nine lives, destiny would dictate that he sacrifice a few more before war's end.

Hindsight is twenty-twenty. Karl's lucky *heimatschuss* had occurred near the thirteenth-century routes traveled by Niccolo and Maffeo Polo, father and uncle of the legendary Marco. They had, some seven centuries earlier, traded jewels and other wares of value between Constantinople (Istanbul), the Crimea, and present-day Volgograd (Stalingrad). After about 1259, the vast area between Constantinople and Mongolia had become what was famously known as the Golden Horde, the northwestern khanate founded by Batu Khan, with his golden tents. We met Batu Khan earlier as the potentate who gave his name to Botoşani, where Karl began his march over territory that would have been familiar to the Polo family. Later, when Karl found himself in Siberia, he was still in the area once under the sway of the Golden Horde.

On the troop train back to Germany, Karl found it increasingly uncomfortable to sit on his wounded leg, and the injured area developed a bump larger than a golf ball. This at least provided him with a visible manifestation of a wound that the doctors could not ignore. At the hospital, X-rays revealed a piece of shrapnel no larger than a matchhead, but it was lodged right up against the ischiadic (or sciatic) nerve, and his leg would not be set right until the shrapnel was removed. This proved difficult, due to the small size of the piece, and the first exploratory efforts to get it were unsuccessful. Eventually, under an X-ray machine, three needles were inserted from different directions to pinpoint the shrapnel, and it was successfully removed.

Karl "called home to tell [his] parents he was in the hospital." His stepmother, he remembers, "nearly died of shock when she heard my voice. It seems my company commander thought I was dead. They had sent all my belongings home with a letter saying I had been killed in action." Fortunately for the Hirsch family and the future of paleontology, rumors of Karl's death had been greatly exaggerated. "The package had reached home before I got to Poland." It had been a case of bad news traveling faster than good news.

Late in January 1943 there was an outbreak of typhus among the wounded, and two of Paulus's remaining generals committed suicide. On January 30, Hitler described Stalingrad as "the greatest battle in our history." In truth it was his greatest defeat. On January 31, he promoted Paulus to field marshal. As no German field marshal had ever surrendered, this was seen as a blatant ploy to have him commit suicide. As it turned out, Paulus had surrendered the same day, giving the Soviets a haul of prisoners that included a field marshal and twenty generals. Hitler was so angry that he vowed he would never create another field marshal, as none could be trusted. Paulus, on the other hand, awarded all his surviving men the Iron Cross for exceptional bravery. The Soviets rounded up the sorry remnants of the Sixth Army, reduced by an order of magnitude from three hundred thousand when first surrounded to a mere thirty thousand men on this day. While the prisoners were promised humane treatment, Soviet compassion extended only so far. Those still on their feet, those who surrendered unconditionally, and those encountered by compassionate captors, under the watchful eye of appropriate supervision, were spared. The Soviets were angry and shot many who showed the slightest resistance or made the slightest of false moves. Several of the most squalid enclaves of near-dead wounded were doused with gasoline and set ablaze.[1]

Despite the discomfort, a wound in the backside had its lighter side, and Karl could grin at the indignity of sitting on a rubber ring and laugh with the doctors who teased him by saying that as an infantryman he need not worry: "Foot soldiers don't need to sit down." After two operations Karl would finally earn his home leave, a chance to kick back and perhaps get drunk as one might expect a 21-year-old soldier to do. In such unsettled times romance was guarded; dalliances with the ladies would not likely lead to serious attachments or promises of rosy futures.

Karl's *heimatschuss* signaled change. Photos taken during a brief convalescence show him in striped pajamas, sitting on a bench, smiling. As a young veteran with a full year of action behind him, he was qualified to be considered for officer's school, and he enrolled on January 21, 1943. In that same

Karl convalescing back in Berlin.

week, officer history was being made in Stalingrad. Hitler was still refusing to let General Paulus surrender, even though the last of the general's ammunition was almost gone and he had twenty thousand unattended wounded sheltering in the rubble of the city, frostbitten, starving, and mostly without weapons. The end of Operation Barbarossa was very near. The Red Army's Operation Uranus had been a success and the early phases of Operation Saturn were underway, promising the centripetal squeezing of the Axis forces back into western Europe, where Karl was already resituated.

Karl, like the Stalingrad survivors honored by Paulus, would be awarded his Iron Cross, his special führer package, due wounded veterans. It was his New Year's present, received on January 26, 1943. Karl's German friends and family, and in later years his American friends and paleontological colleagues, can be thankful Karl survived the death sentence that almost certainly awaited him had he advanced into Stalingrad.

Back in Germany, with needles stuck into his behind to locate and extract the shrapnel souvenir delivered courtesy of Ivan, veteran Karl would be scheduled for a tour of less hazardous duty. From the far east of Hitler's

Karl wears his Iron Cross, awarded in recognition of a
shrapnel wound in his buttock when he was on
the run near Stalingrad.

reach, he would be posted "to the west coast of France to a harbor town called
Saint Nazaire" on the Atlantic Coast at the mouth the Loire River. Declining to undergo the officer training for which he was eligible (and for which
he had initially applied), he was nevertheless promoted to sergeant on the
far western front of the territory occupied by Hitler's Third Reich. French
resistance notwithstanding, this was a relatively safe deployment, in territory that the Germans had not had to fight for when France had capitulated
to occupation early in the war. Karl "lived in a house with a French couple."
Because "the wife was afraid the hobnailed boots . . . would scratch the floor,"

Karl made the soldiers "put felt pads on their boots. . . . From then on she treated me like a son." As a sign of her affection, she would offer Karl snails from her garden, a "delicacy" new to Karl who would remember "one evening . . . talking with her as she wiped one off, snipped off its little intestine, and said, 'Here, Karl, try one.' I thought 'What the hell,' and swallowed it whole. They were delicious."

Karl recalls that "when it came time for me to be reassigned I went to see the Personnel Officer [and] . . . told him I wanted to be a Mess Sergeant." But as a "combat veteran they needed me at the front, [and] I was assigned a heavy machine-gun platoon." So, for three months Karl trained before being sent to Genoa in Italy, where he was assigned a temporary home in one of the coal "bunkers dug into the hill . . . covering the harbor." When one day "two of my boys confiscated a truck . . . we went down to check out the harbor . . . and . . . checked out the warehouses and found one full of wine and cigarettes . . . [and] . . . loaded it all . . . and stored it in a coal bunker." The next day the boys brought him a truck full of Italian lira, "so we had as much money as we wanted." The piracy had been well timed, as only two weeks later "German administrators arrived and commenced to plunder the town," art treasures and all.

Karl's Italian sojourn, practically a holiday, allowed him to fraternize with a girlfriend, fondly remembered as Liliana. She wanted to marry Karl, and "her father, who owned a ship line, said okay but only under two conditions. I would have to become a Catholic . . . and desert the army. . . . Both conditions scared me . . . too much of a free thinker to be a Catholic. She was a very nice girl but not that nice." Besides, "if the army caught me they would hang me." She was nice enough to keep him past curfew on a few occasions and, with the help of a little wine, disorient his bunker homing instincts. Karl would confess to his share of drinking.

In Italy Karl had been nearer to Africa than to the Soviet Union and he re-members meeting one General Erwin Rommel, when Rommel visited Genoa and actually "came to inspect our gun emplacement . . . looking for the officer in charge." "No officer. Just me, Sergeant Hirsch." The general replied, "OK, show me around." This Karl did, unable to tidy the evidence of the "good life" they were living, "lots of cigarettes, liquor, money . . . he saw all of it . . . [and] was a good officer and didn't say anything." Karl "had a good feeling about him because he treated me like a human being. . . . An officer who takes care of his men has men who will take care of him."

Sadly, Karl was in no position to take care of Rommel when the general was in Hitler's crosshairs. Rommel, famously dubbed by British journalists as "the desert fox" for his epic battles as commander of the German Africa Corps, was to become a legend and the subject of many myths. Not least of these was his purported involvement in the 1944 plot to assassinate Hitler and his reluctance to declare himself an antisemite. He had become a hero of multiple campaigns in North Africa, where he was said to have fought as cleanly and honorably as war allowed. The Germans would be defeated in North Africa in spring 1943, and by summer the Allies' Operation Husky was underway to retake Sicily and southern Italy. Rommel was a thorn in Hitler's side, not just because he failed to win final battles but because Hitler could not abide his dashing general's popularity or any hints of disloyalty. Afraid to execute him publicly for his alleged involvement in the assassination plot, Hitler gave him the chance to commit suicide in return for clemency for his family. In avoiding the disgrace of execution as well as making the gesture of sacrifice on behalf of his family, Rommel's legendary status would only grow in the copious postmortem literature that flourished after the war. Karl would tell one more Rommel story, recounting how as the general was conducting another tour, this time with Karl's commanding officer, Karl was dressed down for being out of uniform. Rommel observed quietly and then winked at him.

Karl would later write that "22- and 23-year-olds" were now veterans. He had been statistically lucky in the survival stakes, as "ten of 220 of our old unit were left," the others dead or prisoners. With two three-month seasons behind him in France and Italy, it was back to the Soviet Front to put his machine-gun training into practice.

What crossed Karl's mind with this deployment? Would he feel as Erich Swinger the German soldier-poet had felt when Swinger wrote, "In a few minutes I shall feel under my feet that secret land which I loved as much as I hated it; which satisfied me as no other country did and yet let me hunger as none had done before . . . in whose hot deserts I thirsted and in whose icy tundras I had wept because of the cold . . . that [land] which robbed me of five years of my young life but which repaid me ten-fold with a wealth of experiences."[2]

Military historians would refer to 1942–1943 as World War II's "middle years." Whether Germany and the Axis powers knew the tide had turned against them was perhaps of secondary concern given Hitler's intransigent, and unrealistic, victory-at-all-costs aspirations. But it was not just Rommel and fellow detractors who were a thorn in the side of the Wehrmacht. An extraordinary German named Rudolf Roessler, code-named "Lucy"—a dedicated anti-Nazi operating in Switzerland, where he posed as an innocent bookseller—was to help the Soviet cause so effectively that he was credited with helping bring down the Wehrmacht with intelligence worth "a dozen atom bombs."[3]

Lucy, dubbed one of the greatest spies of all time, effectively proved that not all Germans were aggressors blindly following Hitler. Through ten highly placed Third Reich officer contacts (the Lucy ring), Rudolf Roessler was able to obtain Hitler's latest battle plans and instantly transmit them by shortwave radio to the Soviets—effectively sabotaging, or at least weakening, many phases of Operation Barbarossa. For example, the initial treaty-breaking invasion had originally been scheduled for May 15, 1941, but it was delayed until June 22, when foot soldier Karl would cross the Prut River. The Soviets had been informed of the delay and had had a month to prepare some of the initial resistance that Karl and his comrades faced. Famously, these extraordinary leaks were discovered when Germans captured Soviet command posts only to find the Wehrmacht's plans already in enemy hands! But despite knowing of the leaks, the Germans could do nothing. Lucy evaded detection and ran his successful operation continuously throughout the entire war, from 1939 through 1945. Neither Hitler nor his most loyal generals knew the source of the leaks Lucy used in his fight against Nazism, so the average foot soldier—the average "Karl" or "Fritz," and there were many thousands of each—was oblivious to the invisible forces of darkness and light vying for power over a war-torn world. Fritz and Karl were mere pawns, flotsam and jetsam washed back and forth by the tides of war, never seeing the big picture conveniently drawn up in postwar maps and postmortem analyses. Could a veteran of Barbarossa phase 1 know his days of Italian wine and roses would end with orders to return to the Soviet steppe to resist a rising red tide somewhere between Krivoy Rog and Donetsk, halfway between Romania and Stalingrad.

There was little of Swinger's poetic sentiment for most of the German troops facing the resurgent Red Army in what became known as the Battle of Dnieper. Between August and December of 1943, Ivan had secured the entire east bank of the Dnieper River; liberated Kiev, the Soviet Union's third-largest city; and established beachheads on parts of the Dnieper's west bank. As the Dnieper continued to flow from its ancient source to the Black Sea, eastern Ukraine was again in Soviet hands. As had been the case more than two years previously, when Operation Barbarossa launched across its thousand-mile, north–south front, with the German tide flowing east, the battlefront of the Dnieper occupied a similar longitudinal line, this time with the red tide flowing west and the Axis forces, with Karl again in action, in retreat.

One wonders what Karl would think of the 2022 battle for Ukraine. In a bizarre and topsy-turvy reversal of military history, Russia would become the aggressor, pushing west, while Ukrainians resisted heroically. The 2022 battle would rage over the same territory where Karl had registered millions of forgotten footprints on Ukrainian soil, first in an aggressive eastward advance, then on the defensive as the "Russians" pushed back westward. Karl would perhaps be thankful that war-wary Germans were not again involved in Ukraine, but like all who are cognizant of history he would no doubt lament the futility of war and the unwelcome resurrection of battlefield ghosts.

Karl was rarely prone to such poetic thoughts even if he had sometimes marveled at the wide-open landscapes. His recent sights had been set more narrowly on Italian wine. But the landscape was changing as he left Italy by train for Russia. "We took along cigarettes . . . liquor and beer and we drank a lot. We knew we were going into battle, and we didn't know if we're going to come home." On one occasion Karl was given a bottle containing machine-gun oil. He claims that after taking a gigantic, careless swig, resulting in a well-lubricated bout with diarrhea, he never again got drunk enough to imbibe without carefully sniffing the brew at hand.

Karl's independent-mindedness had made him decide not to train as an officer. It had been his choice to remain one of the foot soldiers, his means of protest against the futility of war. He had at times risked court-martial or disciplinary action. His reluctance may have been a symptom of the turning tides of war, as he would note that the army's administration was "having trouble getting good officers." The good ones "were all dead" and the ones they were getting "had no idea what was going on or how to treat their men." During this second tour of duty on the Russian front, his typical orders were to secure ridges and other strategic points and to use heavy machine-gun fire to hold the enemy at bay during infantry retreats. There were few obvious rewards for being on the losing team. Having advanced a thousand miles from Romania to Stalingrad in little over a year (June 1941–August 1942), there was little glory in being pushed more than halfway back to Barbarossa's starting point.

The day began when Karl's captain ordered him to take his five-man, two-gun unit over a hill and establish a dangerous defensive position. Karl resisted the order and said he would wait, go by night, and not risk the lives of his men unduly. The captain warned him that the penalty for disobeying orders was death, but did not force the issue, and in any case an older, more experienced officer intervened, nevertheless warning Karl that he could easily have been shot. There had perhaps been a faint realization that Karl's concern was for the safety of the men under his command. The incident was reported, and it reminded Karl of the ever-present dangers of insubordination. As others in the Wehrmacht would do, Sergeant Karl Hirsch was in his small way resisting the folly of blindly obeying senseless orders from on high. Futile as the strategy of taking or holding a small hill in the vastness of the Ukrainian steppe might appear, this was day-to-day strategy involving life-or-death gambling for handfuls of men. Meanwhile, through the crackle of shortwave radio Lucy may already have been controlling the movements of multiple armies and tipping the balance of occupation of a hundred hills and valleys. Karl knew the writing was on the wall: "We were retreating from Russia."

From a little unnamed hill Karl was ordered to lead a counterattack on advancing Red Army troops. As platoon leader he took machine guns and mortars and a motley crew of men from various retreating units, with the objective of establishing a fortified post. "There was a heavy fog" when "all of a sudden the fellow I was with and I came across a Russian machine-gun nest. When I turned . . . to see where my other fellows were, I was hit. The bullet passed through my right leg barely missing a couple of hand grenades in my pocket. . . . It struck the lighter . . . in my left pocket, which then exploded. I could put my fist in the hole it left in my leg."

Karl's right boot filled with blood. "I was so bloody and messed up," he recalled, "the fellows I was with thought I was dead. After we had crawled to safety they patched me up. Through the fog appeared two more of our fellows. At first we wanted to shoot them, thinking they might be Russian." Luckily, they did not. The men carried him back to where his unit and supplies were holed up. The base camp supplies provided a half bottle of cognac, which Karl gulped down before he was bundled onto a cart with another wounded man and evacuated to the nearest field hospital. Karl gripped the arm of his wounded comrade so tightly that he turned his bicep blue with bruises. The doctors at first proposed to amputate his leg, but Karl resisted the proposal. "I was wide awake, you see I was scared they would take my leg . . . but I convinced them otherwise. I believe if I had not been a sergeant . . . they would have. Instead, they put it in a cast and sent me to a hospital in Poland. When I realized they were not going to take my leg I passed right out."

It took two weeks to reach the hospital in Poland, with unsanitary hospital stops "infested with bed bugs" along the way. "They would get between my cast and leg and drive me crazy . . . you could see them crawling above you on the ceiling." Finally he arrived, "covered with dirt and blood," not having "bathed in almost a month." A male nurse scrubbed him down, giving him the "best bath" he ever had before putting him "in a bed, a real bed. After a long nap I woke to see what I thought was an angel. I thought I had died and gone to heaven." Fortunately . . . it was a nurse. Once he was back home, Karl decided he never wanted to "return to the front again." Exercising his leg in secret, he "got strong enough to walk without my crutches" but "never let them know how strong my leg really was." Still, his legs had been literally shot out from under him. It was a severe wound and a longer *heimatschuss*. This was hardly the lucky, kick-in-the-pants delivery from death at Stalingrad that his buttock wound had occasioned.

Evacuated to the Berlin area for a month of recuperation, Karl was assigned duty rescuing victims of bombing raids, innocent civilians as well as two of his own men "when a burning building collapsed on them." Images of a desperate mother clinging to a wounded babe in arms were seared into his memory and made him lament that war makes one "like an animal."

"All too soon it was time to be reassigned." Destiny intervened when "they gave me papers for Corporal Hirsch and sent me to a base west of Berlin" instead of the Russian front. When he ran into the master sergeant, he found that "he was an old friend of mine from my rowing days," and he soon learned that the "place was like a people market" where outfits came and "picked men to fill their ranks." One could "even get into the SS," an outcome that could hardly have been further from Karl's war-weary aspirations. Thanks to his friend's advice about how use the must-go-to-the-toilet subterfuge to dodge being drafted for dangerous duty, and also to the confusion involving "another" soldier named Hirsch, our thrice-lucky protagonist, the "real" Karl Hirsch, would be transferred to Doeberitz, just east of Berlin, and not back to the Eastern Front. Karl would "often wonder what happened to the real Corporal Hirsch and if he got sent to the Russian front."

The real Sergeant Hirsch would be promoted to master sergeant by his friend of the same rank. This gave him access to cigarettes and liquor, which he was supposed to divide unequally between officers and common soldiers in a five-to-one ratio. Karl rebelled. When the next "supply truck came in at night my men and I unloaded . . . and divided the supplies up fairly," with no favors to the officers. "So, I got in a lot of trouble. But I didn't care. What's right is right!"

Karl again demonstrated his independent, antiauthoritarian streak and sense of fair play when "one of the women came and told me she was having trouble with one of the officers and wanted to leave." Reacting to this case of sexual harassment, he told the guilty party, a captain, to "leave her alone." The defiant reply was "What are you going to do about it?" Karl said, "I'll show you" and called his master sergeant friend in Berlin "and had her transferred out." While Karl was on the phone negotiating this woman's escape, the captain was eavesdropping and quickly got his revenge. Karl received a three-day sentence for insubordination, which he served in the Frankfurt stockade. This interlude passed in relative comfort, for Karl made friends with the sergeant assigned to escort him to the stockade, and they stopped in to see a cousin in town and procured two hard-to-come-by bottles of cognac, one for the sergeant and the other for the stockade turnkey, who then "treated me like a prince." Perhaps all truly is fair in love and war when cognac is involved.

Not content with getting into hot water with a sleazy captain, Karl managed to upset a general who had ordered him to "build him a shower." This
Karl dutifully did, and to test it he "hopped in and took a good long, hot
shower." When the general came for his turn, "the hot water ran out" and
Karl was "chewed out" for apparently having "the thing put together wrong."
Karl "struggled the whole time he was yelling … not to laugh." But it "was the
straw that broke the camel's back," and Master Sergeant Hirsch was "transferred to an engineering group in east Poland."

These men were defense builders known as "pioneers," a term used for engineers, or sappers, who were often still among the frontline combat troops
and indeed were called *Pioniertruppen* (pioneer troops). Their missions
were often "assault first, and defensive construction second." Readers will
remember, however, that before his compulsory combat deployments, Karl
has served in the Arbeitsdienst auxiliary, digging ditches, farming, and helping with various military activities such as supply and camouflage. As with
so many wartime endeavors, there was a certain futility in building defenses
only to destroy them as the enemy closed in.

It was October 1944, a year since Karl's second wounding and five years
since war had been declared. He had effectively had five deployments: his
Arbeitsdienst apprenticeship; two tours on the Eastern Front, both reward-
ing him with wounds and unwanted shrapnel souvenirs; and brief stints in
France and Italy. His wartime odyssey, however, was not yet finished. His
sixth and last deployment, under the Wehrmacht in Poland, was near Płock
on the Vistula River, one hundred kilometers west of Warsaw. It was on the
way there in the town of Łódź that he saw the Litzmannstadt ghetto, the
second-largest in Poland, through which more than two hundred thousand
Jews had passed, most of them destined for deportation to infamous concentration camps. Karl was seeing the reality of the Jewish ghettos for the
first time and was still oblivious to the worst of the Nazis' large-scale "Final
Solution" extermination plans.

Karl had now twice been deployed close to two of the war's most destructive, deadly, and historic conflagrations. For good measure—or perhaps one
should say "evil measure"—the Nazis accelerated the extermination of the
Litzmannstadt Ghetto Jews as an extension of the Warsaw rampage. When
the Soviets liberated Łódź in January 1945, fewer than nine hundred Jews
were still alive of the two hundred twenty thousand who had once lived there,

The fact that the Russians were closing in on the region by 1944 did not slow down Hitler's and Himmler's murderous campaign of eradication, or "ethnic cleansing," chillingly labeled *judenrein*, "clean of Jews," and more broadly code-named Operation Reinhard. If anything, the Soviet advance had caused the Nazis to accelerate their plan, petulantly, ostensibly in order to eradicate the evidence of the Nazis' genocidal campaign—or worse, to carry out their extreme racist agenda regardless of world opinion. The Red Army was already on the eastern outskirts of Warsaw by August of 1944, and as they regrouped the Nazis planned to destroy the city completely in reprisal for the Polish resistance, famously known as the Warsaw Uprising. The wanton destruction successfully razed almost the entire city, including its libraries and historic buildings, and was undertaken with fiendish and calculated German precision. All this madness went forward despite the Nazis knowing, or because they knew, that the Allies would soon take the city and forever end their dream to rebuild it as an iconic German city. Historians would later describe the total destruction as comparable to that inflicted on Stalingrad.

and fewer than ten thousand survived the Holocaust. Once again, Karl had been all too close to history's darker days and darkest deeds.

The proverbial writing was on any walls still left standing. Płock, another city that had been home to one of Germany's almost three hundred Jewish ghettos, was soon in Soviet hands, forcing the Germans, including Karl, to retreat to Danzig (Gdańsk), once called the "Free City of Danzig," occupying the so-called Polish Corridor or Corridor to the Sea. Historically, Danzig had divided what had been prewar Germany (the Weimar Republic) to the west, and what was East Prussia, effectively a smaller eastern Weimar Republic satellite. In its expansionist years, Hitler's newly minted German or Third Reich had amalgamated both Germanys, Poland, and Czechoslovakia under the Nazi (National Socialist German Workers' Party) banner. But these were Germany's painful contracting years, and Nazi enemies had scores to settle and territory to reclaim on all sides. The Red Army had already overrun East Prussia and much of Poland.

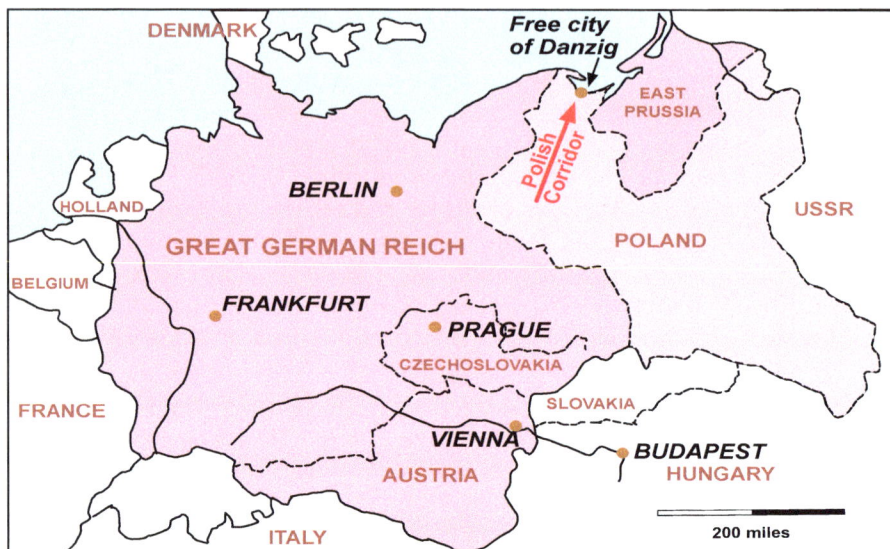

By 1942, the German Third Reich had reached its maximum territorial extent. Poland (stippled area) and the Polish Corridor, where the war had started in September 1939, had been absorbed and removed from the proverbial face of the Axis power's map. Ironically, three years later, as Germany was on the verge of defeat, some of the last desperate battles took place where the war had started, around the once–Free City of Danzig. It was here that Karl, in the wrong place at the wrong time, was captured.

Danzig, where Karl's duties involved blowing up buildings, is the site of one of his most surreal memories. When checking out a manor house, where "we thought there was no one there, we went down into the basement and found a bar" where it appeared there were "three beautiful, well-dressed, nicely made-up women sitting there with an old man. He had his head down as if asleep." But none were moving, and Karl found "the old man had taken a cyanide pill and . . . the three women were just life-sized mannequins." Roguishly Karl would later recall, "We were disappointed, but the liquor helped us get over it."

Karl and his comrades were feeling the final squeeze, as had the old man whose suicide and apparent mannequin fetish would never be known or subject to psychoanalysis. Liquor was clearly a tried-and-true anesthetic. What does one do when the enemy closes in on all sides? A realistic option is to surrender one's hope first, then one's arms and body. Forsaking their duties to lay mines and build defenses, soldiers threw away their guns, stripes, insignia, and pay books. They then hid passively and dejectedly in basements, ready to surrender, and this despite public hangings of deserters at the hands of the remnants of the still-ruthless Nazi terror apparatus.

In some quarters there was little sympathy for fellow soldiers who did not fight to the death. Karl recalled a day when "I saw a truck driving down the street with maybe ten men under guard in the back. When the truck came to a tree with a branch . . . over the street it stopped. One of the guards put a rope around the neck of one of the prisoners . . . the other end to the branch. They hung a sign around the man's neck, 'I did not want to fight for my Führer.' That applied to 'all deserters . . . every one of them . . . hung up all over the city.'"

Karl knew "it was just a matter of time before I was captured or killed." As the "liberators" arrived, Danzig was engulfed in an orgy of angry mayhem. Given the oft-cited heroic resistance of Poles, Russians, and other Soviet fighters against German imperialism, it is all too easy to see the Germans as the aggressive culprits, deserving of the humiliating defeats they suffered. Karl had clearly been on the wrong side of history and had been dealt a bad hand by the gods of luck and fortune. What would become of him now? The answer: "A bunch of us got together in a basement and waited for them. They found us the next day and took us to a prison camp."

German atrocities and Soviet heroism notwithstanding, the Red Army had only so much to be proud of. The history books point incriminating fingers at the atrocities, especially rape, committed by the Soviets and by the Poles in Danzig during late March 1945—ironically the Easter season of rebirth. Karl would vividly recall, along the way to prison camp, "You could hear women screaming . . . being raped. There was a lot of screaming." Karl had recently stood up for a woman's right not to be harassed. As a man he could only cringe inwardly at what he had witnessed. "One thing," he would say, "After the war you would never ask a woman what happened to her . . . it was considered rude."

Pent-up anger against the Germans is often cited as the cause of such frenzied rampages. Although Karl and other defeated soldiers had surrendered or were finally surrendering and German civilians were ready and willing to flee or capitulate to the "liberators," little compassion was shown to enemy captives. Given how the war had started in this same area, this was perhaps not surprising. It was here in the first week of September 1939 that the Germans had committed some of the worst, unprovoked atrocities. Thus, in retaliation, all perceived enemies were rounded up, among them German civilian refugees from East Prussia already in flight before the rapid advance of the Red Army. The Poles systematically attempted to eradicate all traces of German culture and history from Danzig, including the "conversion" of Lutheran churches into Catholic churches. As the city was set on fire, this time by Poles and Russians, the Germans were driven out, and seventy thousand Poles were brought in to replace them. Germans who returned to Danzig were forbidden to speak German, and indeed the city was renamed Gdańsk. Retribution for Nazi ethnic cleansing was swift and decisive. In short, a once-free city, once richly proud of its 90 percent German population and heritage, would be rebranded an "eternally Polish city" in what had long been the strategic Polish Corridor and, in a stark turning of the tables, cleansed of its Teutonic history by Slavic and Communist decree.

Karl had played no part in the 1939 invasion of Poland, regarded universally as the trigger that had started the war some six years earlier. However, this did not inoculate him from being in the wrong place at the wrong time and running into the virulent retribution that ravaged this historic, twice-contended battleground. It was out of the Ukrainian and German frying pan into the Danzig fire.

Simpler histories tell us that England and France had appeased Hitler by failing to resist his annexing of Czechoslovakia in 1938. However, they had warned him that any invasion of Poland would be a step too far. Hitler countered that the area around the free city of Danzig, created in 1920 as part of the Treaty of Versailles at the end of World War I, and situated in the Polish Corridor, between what was west Germany and East Prussia, had traditionally been "ethnically" German or Prussian territory. Indeed, in a frenzied, sweaty Berlin speech on September 1, Hitler claimed that "Danzig was and is a German City. . . . The Corridor was and is German," previously saved by Germany from "the deepest barbarism."[4] Hypocritically he ordered one the most barbarous invasions of the war, on a zone designated as "free by international treaty." His barbarism beyond preemptive artillery bombardment included the rounding up, imprisonment, or murder by the Gestapo of Polish officials, teachers, and priests and the issuing of orders for the euthanasia—that is, the murder—of some seventy thousand of the physically and mentally disabled.

In 1939, Hitler believed that in the war of nerves, Britain and France did not have the stomach to stand up to him if he "only" invaded this small part of Poland in what he misleadingly suggested could be a small, local war. It was clearly too late for Britain or France to hope that their requests for a German withdrawal from Poland would be heeded or to have any hope of averting war.

This was the dark history that hung over the once–free city of Danzig as Hitler launched his 1939 invasion and Britain and France retaliated by declaring war on Germany and declaring their support for Poland. Six years later, with historical hindsight, deepest barbarism would come home to roost. With grotesque irony and many bitter doses of retributive justice, any German soldier captured in the Polish Corridor would answer harshly for Hitler's "senseless ambition" and many war crimes.

The last German soldiers who had not been captured were driven from Danzig (Gdańsk) by March 29, 1945. Karl was not among them. The face of Europe was changing, and its ever-shifting national boundaries and allegiances were again in flux, as they had been since the outbreak of hostilities. Had Karl not been squeezed into the notorious corridor as the war was about to end, he might have escaped capture by the Soviets and the retribution that followed.

Notes

1. The heroic success of the Soviet Army in defending Stalingrad against the German assault, and the appalling conditions faced by German soldiers caught when their supply lines were cut off, has been retold in countless books and reports on this pivotal turning point in the war.

2. Erich Swinger's poem is cited by J. Lucas in *War on the Eastern Front: The German Soldier in Russia 1941–1945* (London: Greenhill Books, 1991).

3. A. Read and D. Fisher, *Operation Lucy: Most Secret Spy Ring of the Second World War* (New York: Coward, McCann & Geoghegan, 1981).

4. From Adolf Hitler's Declaration of War, September 1, 1939, Reichstag, Berlin, Germany.

CHAPTER SIX

Siberia

> The Russians think differently.
> *Karl Hirsch*

The Soviet officer was drunk and seemed preoccupied with the spinning top. He played with it on his desk as if he had never seen one before. But he had two other things on his mind: the minefields and Karl's boots. Of course, it was his job to interrogate prisoners about the location of minefields, and the Germans had laid plenty of mines around Danzig. "I told him all I did was dig ditches. I convinced him I was just a plain old soldier and he let me back with the other soldiers." The fact that "he was dead drunk" is probably "why he let me off so easy."[1]

Though the German prisoners had all thrown away their papers, Karl's good leather boots made him appear like an officer masquerading as an enlisted man. In a sense, Karl was not lying when he swore that he was a foot soldier with no rank. The drunk Soviet officer probably did not believe the story, but he hardly cared. When Karl obeyed the order to roll up his sleeve, the officer had found no blood-group tattoo. Clearly Karl was not a member of the SS, nor was he a high-ranking officer. He had probably procured the boots somewhere along the way and would soon be wearing them no more. For all the officer cared, he could be sent off with the other prisoners to face a future without rank, privilege, or decent boots. In fact, upon "marching to the camp a Russian soldier waved me into a house. He wanted my boots . . . and he gave me his, which were too small. He said cut out the toes."

East of Danzig lies the Frisches Haff, or "fresh lake," an elongate lagoon, three or four miles wide and fifty miles long, separated from the Baltic Sea by barrier islands. Some four centuries earlier, Canon Nicolaus Koppernigk, better known by his Latin name Copernicus, had lived in an isolated tower at Frauenburg on the shores of the Haff. There, beside what he regarded as depressing vapors and exhalations from these waters, he gazed apathetically at the night sky while his seemingly dejected brain reformulated a heliocentric model of the solar system. The sun had not been at the center of the universe since the enlightened days of the Greek astronomer and mathematician Aristarchus, in the third century BCE. For more than a millennium, an irrational, half-religious dogma had banished the sun from its rightful position. Yet it was at the end of this protracted age of darkness, amid brutal warfare between the Polish king and the degenerate Order of Teutonic Knights, deep in a gloomy mind, in a gloomy tower, that the sun began to shine again in its rightful place in the center of our solar system. Nicolaus Koppernigk would never dream that in centuries to come, as the balance of power swayed between east and west, battles would be fought in which military operations would assume the names of the planets in his heliocentric system of astronomy. The Red Army's Operation Uranus and Operation Saturn had resoundingly pushed the Germans far back from their eastern-most advances. Symbolically the centripetal squeeze on the Axis powers had long ago constricted their movements to the inner orbits of their dwindling power. With the Axis collapsing ever inward, there had already been at least two iterations of Operation Jupiter exerting pressure on the Axis forces from Scandinavia and France. The Allies had the inner orbits of Axis power entirely surrounded and ready to implode.

After five years of brutal warfare, the sun was reluctant to shine roundly on the affairs of men. Yet, like Copernicus, compromised by gloom, and desensitized by war, the light of life was not entirely extinguished in the dejected hearts of Karl and his fellow captives. There were still many miles to walk, but at the end of the trail the war might end, and life might resume.

It was March 27, 1945, and Karl had no way of knowing that for some, the war would end within six weeks. For others, including Karl, it would drag on for years. Most German prisoners were held in Danzig for no more than a week. "They shaved our heads and deloused us . . . and broke us up into groups of one hundred." They were then herded onto a train that trundled them east out of Danzig; there they were unloaded and forced to march to a railhead in East Prussia.

The prisoners marched east, hungry and exhausted, sleeping out and trying to help wounded comrades. Those who fell by the wayside were unceremoniously shot and replaced by local Polish people. The Soviets were obsessed with quotas, and each unit had to contain a hundred men, regardless of where they came from. "If the group started out with a hundred it had to end with a hundred or the guards got in trouble." Each cattle car on the railroad truck was also crammed with a hundred men, regardless of the size. Karl and his sorry contingent were crammed into a small car, in which it was impossible to lie down. A hole was cut in the floor "that you could do your business through. We stayed in that car for about two weeks until we came to the camp at Nishne in Siberia."

After the treatment the Poles had suffered at the hands of the Germans, they had been happy to cooperate with the Soviets in opening the way for the counteroffensives that were now leading to the conquest of Germany. It was therefore unjust to press-gang the locals into labor camps for no apparent reason other than to make up quotas, when wounded or sick prisoners fell or were killed along the way. But they treated their own men the same way, and countless Soviet soldiers ended up in Soviet labor camps once they were released by the enemy. As Karl was to often remind himself, "The Russians think differently."

The long train journeys and marches eventually ended at Nishny Tagil, near Sverdlovsk (now called Yekaterinburg), Siberia, east of the Urals. These were huge industrial areas, where the Russians had relocated much industry in the early years of the war. It had not been uncommon to move a tank factory or rolling stock factory east of the Urals, lay out a floor, and begin manufacture of products before the facility's walls and roof had been built.

The prison camp barracks were half buried with an embankment all around; the beds were bare boards on which prisoners slept with no bedding other than a rag under one's hip and shoulder. There were no fittings in the barracks other than a stove, for which little or no fuel was provided. "There was never enough to eat. The guards did not get enough food to feed themselves so of course they stole ours." Washing facilities consisted of a crude type of outdoor shower where inmates could wash out their meager foot rags. No one owned any personal possessions other than items made from scraps of clothing. The few books that were available consisted of the propaganda of Marx and Lenin.

The prisoners had to collect their own firewood when working in the fields and smuggle it "home" to the barracks by hiding it in their clothing. They did the same with seeds of all kinds, the heads of wheat plants, and other miniscule food items. All such furtive collecting was technically illegal and regarded as theft of communal property. If prisoners were caught, the guards could punish them, but such punishments were not usually severe and could sometimes be comical. The guards might take a lazy swing at the offender, who would appear suitably contrite and humiliated. For all that, it was no picnic. "We worked in the heat of summer and the bitter cold of winter."

Some prisoners attempted to escape, and on one occasion two men got away for five days. After they were captured and returned to the camp, their offense was broadcast on the public address system. The propaganda authorities outlined that, under the Geneva Convention, prisoners were expected to try to escape, and that for this reason there would, supposedly, be no punishment for the escape attempt. The offenders had stolen food and clothing in their escapade, however, and for this a severe punishment was warranted. Both were sentenced to ten years in the salt mines. "Why shoot them when you can work them to death."

Most prisoners worked on some type of collective farm, or *kolkhoz*, where labor consisted of harvesting potatoes, cabbages, and other vegetables. Such work gave prisoners the opportunity to supplement their diets with raw vegetables. They also collected nettles, dandelions, pine needles, and other plants for vitamins and to makes teas. Many prisoners suffered night blindness as a result of undernourishment, and most found their teeth deteriorating. In short, many were suffering from tooth-rotting scurvy. With no hope of dental treatment prisoners developed cavities, which they would sometimes stuff with nicotine residue from tobacco pipes in order to alleviate toothache and infection.

Stories of escapes by individuals and groups of prisoners from Siberian prison camps during World War II have made for thrilling popular literature. Classics include *As Far as My Feet Will Carry Me*, the story, made into a book and a film, of one Clemens Forell, a German soldier who claimed to have escaped from hard labor at a lead mine on the far-east Chukchi Peninsula and walked and hopped trains, with help from a dog and Siberian reindeer herders, finding his way to Iran over a period of more than a year. Equally famous, and memorialized by the book and British radio series, is *The Long Walk*, the story of Polish soldier Sławomir Rawicz, who escaped with six other men from a snowbound Siberian gulag and walked into Mongolia and across both the Gobi Desert and the Himalayas to salvation in India. Only half the group survived the ordeal. Both stories were subsequently ghost-written, for men with poor literary skills who had wandered, sometimes starving and delirious, in the wilderness with few geographical guideposts. While the protagonists appear to have been real, subsequent research has revealed flaws and inaccuracies in the accounts. The stories nevertheless make for compelling reading.[2]

"For some reason," Karl recalled, "the Russians let me keep my pipe, and you could get tobacco easier than cigarettes . . . the guards . . . and even the camp commander . . . would call me in so they could use my pipe . . . and usually gave me a little tobacco."

Almost everything worked on the quota system and meeting a quota was the incentive for earning one's essential food ration. Workers went out in groups of one hundred with quotas to fill. When harvesting potatoes, for example, workers were given a set time to harvest a particular area. This encouraged inefficiency, and much of the crop was left in the ground. In theory there was a reward system for completing one's assigned chores. Prisoners were paid with tobacco, not cash, but there was often a catch. In order to get paid, one had to work a complete week on a particular job; but more often than not, on the sixth or seventh day workers were reassigned to new tasks.

Karl and his fellow prisoners grew resigned to the relentless drudgery, and most of them made some sort of adjustment to the meager diet and backbreaking work. After the first snows, the cabbages were harvested in large baskets the size of bathtubs and brought in on horse sleds; this was the time to catch a moment's relaxation, and be carried drowsily along, listening to the sound of jingling harness and sleigh bells.

Exhaustion and hunger were always the labor camp inmates' close companions. Food was always on everyone's mind, though some like Karl who were lean and mean survived better than others on the near-starvation diet. Others became obsessed with the shortages, and this just made it worse. The guards were also poorly fed and often stole food from the pitiful supply allocated to their prisoners. Kitchen duty offered the opportunity to pilfer raw fish or some little extra scrap of food. The typical diet was a soup or gruel rarely containing more than cabbage, or a little fish and a moist black bread. Inmates were always hungry, and it seemed they had to stand in line endlessly to obtain what little rations they were given. After the war Karl never again felt comfortable standing in line for food. There was so little spare food that it seemed one never saw a mouse or a rat around camp. The rodent tribe also went hungry.

Stealing food from the kitchen was a crime in the eyes of fellow prisoners, as well as in the eyes of the guards, so it was important to exercise discretion in any pilfering that might deprive others of their share. But a far greater crime was the theft of food from a person's private store. Anyone caught stealing a ration that had been set aside or saved for later was immediately subject to barrack room punishment. This invariably involved a severe and humiliating beating by other inmates. The offender would be stripped and lashed with belts and fists, until suitably bloodied and thoroughly deterred from any thought of further theft.

Poor diet and backbreaking labor resulted in sickness and exhaustion and could qualify an inmate for a period of recuperation in the hospital ward. But the line between work and recuperation, or health and sickness, in such circumstances was tenuous—a line between life and death. It did not pay to be sick for too long and lose one's usefulness as a laborer. Losing one's will to recuperate and survive could be the beginning of the end, for it made a person expendable in the eyes of his captors.

Those with spirit and will could use a stay in hospital to regain a little strength and perhaps pilfer some item of material or food that they might not otherwise have access to. Staff were wise to such pilfering and would spring regular inspections on the inmates to make sure they were not removing material from the bedsheets or the furniture to trade for food or tobacco. Most inmates made makeshift knives from pieces of metal or fashioned teacups and coffee cups from tin cans, and some even made rings from brass nuts. It was common practice to cut a strip of material from the hemline of coats and other garments and use the material to make wallets or pouches to trade for tobacco and other meager items.

Agricultural labor was seasonal, of course, and it was impossible to benefit from the slim pickings of the harvest season during the rest of the year. Inmates were therefore assigned all manner of other forms of physical labor, including construction of ammunition boxes, unloading of railroad freight, factory work, lumberjacking, and snow-shoveling. Much of it was backbreaking work. Trees were felled using a two-man handsaw, and an eight-hour quota often took ten or twelve hours to fill. In winter the men were excused from work only if the temperature fell below −45°F. The snow was often more than a meter deep, and shoveling it away required cutting it down like a staircase one step at a time. In his postwar life this was a chore that Karl could enjoy, for he knew that he was doing it for himself and that if he wanted to he could always stop and "go have a cup of coffee." We have shoveled snow with Karl in Colorado and seen him grinning happily at the thought that he was working voluntarily and on his own schedule.

The heavy work took its toll on the inmate population. It is hard to say how many men perished there, but in a given year it could have been close to half the population. Many froze to death; others couldn't sleep and so became exhausted and sick. All were emaciated, but deep within himself Karl knew he was strong, and that in these tough circumstances his slight but wiry build was more of an asset than a liability. On occasions when he began to suffer from diarrhea, he stuffed himself up by eating charcoal obtained from wood ash. It would not do to dehydrate through sickness. His weight was already down to a mere one hundred pounds.

The trick was to "roll with the punches," and in any spare time that was available it was best to just sleep off the exhaustion. There was also a simple trick to taking a break, and that was simply to smoke. Nonsmokers were discriminated against and could not stop for a breather without the guards ordering them back to work. But it was acceptable for smokers to take a break. On November 7, inmates were given a shot of vodka to celebrate the Russian Revolution. Such celebrations were accompanied by searches of the barracks. The Soviets evidently feared that the alcohol might encourage inmates to run amok.

The history books tell us that the war ended on May 8, 1945, but before this date the Germans already knew that the end was approaching. The prison camp's public address system announced the success of the Allies and the imminent conquest of the once-powerful German war machine. Rumors arose that one could escape the camps by going to fight to with the Russians against the Japanese. But for many inmates, life continued unchanged. History also tells us that some prisoners were not released until 1955. Those who had special skills were kept on as long as they proved useful. The sick and infirm were often released first, but many were in such poor condition that they did not make it. Again, there was a fine line between being sick enough to be released and being healthy enough to be kept on as useful labor.

Various postcards addressed to Karl dated between September 1945 and November 1947 have survived, as well as a handful that Karl was able to send sporadically to his family and friends. A number of cards dating from 1946 and addressed through the prison camp system and its censors, including one from sister Anni, show that Karl noted the arrival dates, sometimes up to seven months after the cards were sent. They evidently traveled much more slowly if mailed via Moscow in midwinter! One longer letter survives from the last few months of Karl's captivity. It was sent by his half-brother Hans Jürgen on March 15, 1947, five days before Karl's twenty-sixth birthday.[3] It too was addressed to the central prisoner-of-war (POW) clearinghouse (the "Kreigsgefangener") in Moscow, but it did not arrive until May 4. On it Hans wishes Karl a happy birthday and recognizes that he is again celebrating "far from home." News of the family reported by Hans Jürgen reflects the postwar confusion. The card is not clear on where all family members are, or what they might be doing, as displaced persons remained on the move, including "a steamboat with Germans arrived from Japan." These were German prisoners whom the Soviets had cunningly "requisitioned" to fight the Japanese!

Hans Jürgen asked if Karl had received other letters, from sister Anni, for example, and reports that he has not yet found work. Hans had also been a POW and reported that he suffered from rheumatism, but a photo he had

СОЮЗ ОБЩЕСТВ КРАСНОГО КРЕСТА и КРАСНОГО ПОЛУМЕСЯЦА
СССР

Почтовая карточка военнопленн~
Carte postale au prisonnier de guerre

Кому   (Destinataire) *Karl Franz Hirsch*

Куда (Adresse) *Rotes Kreuz   Moskau*
(страна, город, улица, № дома, округ, село, деревня)

Postf. 245/2

Отправитель (Expéditeur)
Фамилия и имя отправителя *Claus Groth,*
Nom de l'expéditeur

Почтовый адрес отправителя *Berlin-Halensee,*
Adresse de l'expéditeur *Joachim-Friedrich str. 7 b. klener*

Prière d'écrire sur carte postale, autrement ces lettres ne seront pas remises au destinataire.
Lettre au verso.

One of a dozen surviving letters and postcards that Karl received or sent while he was imprisoned in the Soviet Union for two and half years. This one was sent from Berlin on December 29, 1946. Karl noted the date it was received: February 9, 1947. Some deliveries took much longer, up to five months.

enclosed shows that his face had filled out well. It appears that the family may have known that Karl was still alive, but two years after war's end, as time dragged on, one could not be certain. His father added a footnote to Hans's letter, wishing Karl the best for "your new year in life" and hoping that Karl has survived the hard winter and might come home by summer. This finally would come to pass.

Karl's chance for freedom came, as he wrote later, when he was in recovery barracks and "heard a rumor . . . that they had trains going back to Germany" with priority for those too weak to work. "I began starving myself so I would stay weak." He knew he was "walking a knife edge." Too weak and you might die, too strong and it was back to work. Years later he would not remember exactly "when I found out I was going home. One day I was put on a train for Germany. It was the cattle cars again." The ride west took ten days, with the wagons open. At stations along the way, women were selling stuff and sometimes gave the men food. It was August 1947 and Karl "had been prisoner for two and half years"—and the war had been over for most of that period.

His liberation had been on hold as the Soviets extracted their pound of German flesh from the sweat of his brow and the atrophy of his undernourished muscle tissue. They were indifferent to his future aspirations. He would later write the eternal geological truth: "Ein steter tropfen höltden stein"—a steady drop hollows the stone.

Notes

1. Unless otherwise stated, all quotations in this chapter that are obviously in Karl Hirsch's voice come from his brief biography *Walking in Your Moccasins* (WIYM), from the Hirsch Audio Archive (HAA), or from notes made by the authors when interviewing Karl directly.

2. J. M. Bauer's book *As Far as My Feet Will Take Me* (New York: Random House, 1957) is the story of a German prisoner, Clemens Forell, who escaped from a Siberian prison camp and walked to Iran. The 1955 book *The Long Walk*, by Sławomir Rawicz and Ronald Downing, (Guilford, CT: Lyons Press, repr. ed., 2016) is the story of a Polish soldier, Sławomir Rawicz, who escaped with five others from a Siberian camp in the Arctic Circle and walked to India. The authenticity of both stories has been challenged.

3. Copies of some of the postcards and short letters, written in German, to and from Karl and his family have survived and are part of the archive used in the preparation of this biography. See the example in chapter 6.

CHAPTER SEVEN

Second Birthday

*The look on her face when she saw me was something
I will never forget. It was a look of disbelief and joy at the same time....
Everyone thought I had been killed.*

Karl Hirsch

Karl's first destination upon leaving the POW camp was the town of Frankfurt—not the large city in southern Germany, but the town on the "Oder line" that divided Poland and East Germany. Recall how Hitler dreamed he was about to be engulfed in a trench collapse and only just got clear when an enemy shell hit and killed all his companions. Hitler interpreted this dream as an act of Providence intervening to save him for a greater destiny. Karl too was about to be saved, and he would remember August 2 as his "second birthday."

The released captives were deloused and fattened up for a week or so, with a little better food. They were even given better clothes, in Karl's case a Russian coat and wooden-soled shoes. He could have passed for a Russian. Then, for the first time in years, they asked him where he wanted to go. Karl did not write much about his experiences until years later, and then not in great detail. However, he would relate this second, life-changing *heimatschuss* or "home shot" in the simplest of words:

> From Frankfurt I went to Berlin. That was where my family had lived before the war. I did not know how much of my family was still alive. I remembered that my sister, [Anni] a nurse, had worked in a hospital in what was then East Berlin. I called there and a woman told me that my sister was in Hamburg. That worried me because the

last I knew Hamburg had been a mental hospital. I asked the woman if my sister
had gone crazy. She said that Hamburg was a children's hospital and that my sister
worked as a nurse. I called my sister and she told me that everyone was at home. [As
Karl would relate, Anni wept tears of joy.] My parents and brother who had also been
a POW had made it home before me. My rich aunt and her husband. Even my communist friend who had been a POW in the US was there.

   I took the streetcar which at the time still ran into West Berlin, to my parents'
house. They lived up one flight of stairs. I had to stop in the middle and rest because
I was still pretty weak. When I got to the top . . . I sat down to catch my breath. My
mother heard some noise in the hall and came to see what it was. The look on her face
when she saw me was something I will never forget. It was a look of disbelief and joy at
the same time. . . . Everyone thought I had been killed.[1]

Karl stayed with his parents and slowly got his strength back. Anni brought
food from the hospital, one of the few places with extra food. He soon went
to see his Communist friend, who had been thrown in jail by the British.
He remembered his friend's mother as someone he could confide in when
he was younger, so he told her some of the terrible things he had seen. She
didn't believe him, saying "Communists would never do such awful things."

It was still difficult for people to get the things they needed. His rich aunt
had managed to save most of the family's bicycles, though not Karl's, which
they claimed the Russians had stolen. Karl suspected the family may have
sold or traded it to obtain food. But quite possibly his relatives were right:
from the time the Soviets invaded and "liberated" Berlin, the average foot
soldier was like a proverbial kid in a candy store. Many a provincial peasant
Red Army soldier from some the remote steppe had never seen a bicycle
before the war, and certainly had never owned one. As the anonymous German author of the classic *A Woman in Berlin* (later identified as the German
journalist Marta Hillers) would recall, it was comical to see young men learning to ride a bike. Stealing bikes was one of the less egregious activities the
Soviets engaged in once they occupied Berlin in 1945. Berlin residents had
had a two-year head start on Karl in observing Soviet behavior and honing
their survival schemes and strategies.

It was difficult to get a job in Berlin, even though Karl had obtained letters
of recommendation from both sides of the political spectrum: his Communist friend and a Christian Democrat. Technically Karl had never been
ousted from the Postal Service, where he had been employed a decade previously. But when he went there to apply for work, there was little sympathy

for ex-prisoners. "I probably blew my stack," he recalled, seeing the postal employees as fat cats, and seething as he told them "You should have been in Russia." Considering what he had been through, Karl rarely expressed anger, but natural resentment would bubble up when he encountered petty injustice.

Undeterred, Karl was able to reestablish his presence in the community. He obtained a post working with a Youth Education Organization, which was part of the Berlin city government's welfare system, composed of youth houses ("Jung Haus") distributed around Berlin's various city districts. It was here in a tent camp named Düppel that Karl met Gerhard—who was later to become his brother-in-law. Gerhard was in charge of one of the youth houses and had the job of distributing a limited store of supplies, some of which were donated by American troops. In these postwar years everything was in short supply, and suspicions, squabbles, and differences arose easily, especially over the acquisition and distribution of food and fuel. Karl even had to separate adolescent boys and girls who were sleeping in the same tents. They objected, of course, but their attempts to besmirch no-nonsense Karl were petty and of little or no avail. All the trees had been cut down for firewood, and during the winter of 1947 the planks on the footbridge over the Spree River began to disappear. At first, every other plank was removed, but by spring only every fourth or fifth plank remained, leaving just enough support for people to hold on to the side rails and step across. A carton of cigarettes could be traded for a week's worth of food.

Gerhard's sister Hildegard was living in the village of Ruhlsdorf in East Germany when she met Karl at a Christmas party at the youth center in Berlin. At the time, she was working as a secretary at a pig farm research center, already in the eastern sector under Russian control. The Berlin Wall had not yet been erected, but there were many checkpoints separating the eastern Communist sector from the southwestern British and America sector and the northwestern French sector. There was a rather slow, impeded osmosis of people in both directions. The Communists had it in for Hildegard's boss, a Catholic, and wanted to frame him on a bogus charge. They wanted her to speak out against him, but she wouldn't and so was threatened with jail. A sympathetic local policeman suggested that she "get out while the getting's good," so she grabbed her bike, her dog Bingo, and her knapsack and fled west. Before the wall was erected, one could still escape through houses, subways, and sewers.

"Safely" in West Berlin, she went to live with her mother, where she was twice investigated by plainclothes police from the east where she was still scheduled to appear in court for the phony case against her boss. On the run, she went to live with Karl, working at whatever jobs she could find.

Bingo was a German Shepherd that Hildegard had trained herself. He was very loyal and protective, growling and baring his teeth the first time Karl put his arm around Hildegard. Hildegard forced Karl to teach Bingo who was boss, and from then on Bingo relented. Dogs remained part of the family, with Hildegard specializing in dog training at a kennel club. Times were still difficult in Berlin, but the couple enjoyed a little sparse domesticity in their small apartment. They obtained a few pieces of silverware and crockery on the black market. Not surprisingly, there was little love lost between the Germans and the Communist Russians, the beaten Germans sometime itching to fight back when abused by the Communist authorities. The Russians could jail citizens just for talking back and could dish out harsher, even lethal, punishment for more serious infractions. On one occasion Karl was carrying a cake made by Hildegard as he passed a checkpoint. The soldiers took the cake, ate it in front of him, and jeeringly thanked him for the "gift." Karl would write, "If I had a gun, I would have killed them all." Not for the first time, he thought of escape and a new life.

Karl sometimes faced angry mobs who accused him of stealing incoming supplies, especially meat and cigarettes, and selling them for profit. He dealt with these confrontations by offering to let the youths come and inspect the inventory when it was delivered. After all, even in prison camp there had been a stern code of ethics. The kids were tough and streetwise, and though they were aware of the moral values expounded by pastors and other leaders who were trying to rebuild the community, they were also realistic if not half-desperate. They argued that they had had to steal to feed their parents and themselves, and they had to survive any way they could. Karl saw the same tough breed when he was transferred from the youth house to a city-run refugee camp. Here refugees from the east, euphemistically called "displaced persons" (DPs), could make five marks a day cleaning and recycling bricks from bombed-out buildings. On one occasion Karl did not have the keys to open the office safe, but he soon found a young fellow with the requisite lock-picking skills.

The rough-and-ready kids did not always impress visiting dignitaries, and on one occasion an American general who had come to visit the summer camp complained that the kids were unruly and did not stand to attention or salute. This complaint came to nothing when other, more realistic American visitors turned in favorable reports.

In these hard times it was easy to suspect anyone of theft and of grabbing what they could whenever the opportunity arose. So it was that both Karl and Gerhard came under suspicion of embezzling supplies and were investigated by city officials. The allegations were that they had passed on supplies of coal and firewood to friends and relatives. This was true only insofar as they had rewarded friends and family who had come to these centers, sometimes for long hours, to help out with day-to-day operations. Karl warned his detractors that he and Gerhard had never once sold a single item, or violated any reasonable principle of ethics, and that if he was charged with any wrongdoing he would fight the case with all means at his disposal. There were six hundred people in the refugee camp, all poor, deprived, and on the edge of desperation. Cynics have said that pettiness raises its head when so little is at stake, but what was at stake was the future welfare of each and every displaced person.

Karl had already started his journey west when he was trundled two thousand miles by cattle car, from east of the Urals back to Frankfurt, pausing only briefly before the last fifty-mile leg to Berlin. Symbolically Hildegard made her much shorter journey west from the eastern Russian sector into West Berlin. For the next two generations, the Western world would recognize the stark political division between East and West Berlin: the infamous "Berlin Wall" and the Cold War chill that defined the division between the Soviet Eastern Bloc and the Western Bloc administered by the United States and Western allies.

The first half of the twentieth century had seen two devastating world wars, in which Germany had played notorious roles. The second war had thrown Karl into the maelstrom, mauling his legs and psyche and leaving him lucky to survive. He was twenty-nine years old, and the last decade had

Historians and political scientists have talked about a century-long cycle of four generations when the mood of society changes from crisis (winter) to rebirth (spring), awakening (summer), unravelling (fall/autumn), and back to crisis. Each season has a birth generation, which grows in subsequent seasons into young adulthood, middle age, retirement, and old age. The crises of depression and World War II, roughly dated between the 1920s and late 1940s, had bred a generation of young adult "heroes" destined to try to fix the crises. In the 1950s, the season was turning from winter to spring, from a generation of heroes to a generation that would give birth to the baby boomers. By these generational theory definitions, Karl and Hildegard were, as young adults, of the crisis-weathering, hero generation and were and ready for spring and new life in their middle years.[2]

been frequently traumatic, largely wasted, and physically and psychologically challenging throughout. It was therefore symbolic of a new postcrisis mood that he and Hildegard would get married in January 1950, as the second half of the century dawned. Karl would write simply that he "felt the time was right to ask Hildegard to marry me." She agreed and they had a simple wedding, with Karl dressed in a new suit donated by a church group. It was the first time he had had "new clothes . . . in a long, long time."

Even before their wedding, Karl and Hildegard had decided that they "wanted to leave Germany and start a new life." The couple applied to emigrate to Australia, Canada, and the United States and vowed that wherever they went, they would never return, even if they never rose above the menial status of janitor. The prospects in Australia did not look good; men and women were often separated and sent to work in different locations. Emigrants to Canada could not easily rise beyond immigrant status. In the United States, however, one could aspire to citizenship and a new life of new opportunities.

Because Karl and his family originated in the Landsberg area, east of Berlin, he qualified for assistance under a program for displaced persons. Hildegard had also been displaced. Through his contacts at the Youth Organization,

Ú. S. NAVAL SHIP GEN. C. C. BALLOU

In 1952, the ten-thousand-ton American "refugee ship" brought long-suffering Karl and wife Hildegard from Bremen, Germany, to New York, where they would begin a new life in a new world. Used with permission from Navsource.org.

Karl met a Hungarian minister from Budapest who went on speaking tours in the US and met families there who wanted to adopt immigrants. Karl had helped the pastor with fifty pounds of coal (decades later he still had the written IOU). Miraculously, Karl and Hildegard were on their way to the US by summer of the same year, 1952.

Leaving from Bremen, where they had worked in an embarkation camp, they sailed west across the Atlantic to New York in a ten-thousand-ton troop transport ship named the *General C. C. Ballou*, which was pressed into service for almost two years as a refugee ship.[3]

The fourteen hundred men and women were kept separate and helped pay their way by painting the decks. For three days they weathered a gale and were confined below deck. Postwar tensions still simmered, and upon arrival in the US, dock workers, not known for refinement or tact, taunted them with

The Hirsch trunk, which held virtually all their worldly possessions, accompanied Karl and Hildegard from Bremen, Germany, across the Atlantic and the Appalachian Mountains, to their temporary home in Port Allegany, Pennsylvania, known historically as a canoe portage and pioneer route site used by westbound migrants. Karl and Hildegard would continue this tradition and eventually haul their trunk to the Wild West.

cries of "Nazi." They set foot on American soil on June 10, 1952. Karl was 31 and Hildegard was 28.

Karl and Hildegard brought with them a single trunk stenciled with Karl's name and the couple's destination, "PORT ALLEGANY, PENNA U.S.," a small town situated near the source of the Allegheny River, which flows from north-central Pennsylvania three hundred miles to the southwest corner of the state and its confluence with the Ohio River at Pittsburgh. Port Allegany, meaning "Canoe Place," is not a port in the traditional sense. Situated in the sparsely populated Pennsylvania Wilds on the east side of the so-called great divide of the Appalachian Mountains, it was a portage place some twenty miles west of the Susquehanna River where Native Americans camped and made canoes for their westward journeys. Also heading west, with the great Appalachian divide behind them, Karl and Hildegard would trade a sojourn on their ten-thousand-ton refugee ship for a temporary portage home at the Canoe Place.

Karl chuckled about the bad language of their Hungarian minister friend, who despite his profession "cussed a lot." But he was "a very smart man" who had fulfilled his savior role by finding a US family who would sponsor the displaced couple. This was arranged through the Presbyterian Church, allowing Karl and Hildegard to emigrate to Port Allegeny, Pennsylvania, again heading inexorably west, two hundred miles beyond New York City. They moved initially into a Swedish and Norwegian community. Germans at first had immigrant status, a low rung on the ladder of upward mobility. For two years Karl worked as an unskilled laborer in the Pittsburgh Corning factory, manufacturing glass blocks, and at times he had opportunities to do remodeling and gardening work on the side.

After his first week in Pennsylvania Karl's leg swelled up, perhaps from using new muscles, and his shrapnel-bespattered leg became infected. The doctor had to cut into certain areas to drain the infection. Karl remembers coming in from the garden and going to his factory job with dirt still under his fingernails. He had been exposed to much squalor in the previous decade, but this did not mean a doctor should ignore simple hygienic procedures. After the treatment the doctor got out whiskey for Karl—and also gave some to Karl's sponsor, who was faint at the sight of the bloody wound.

So, Karl and Hildegard began to become American. They learned to eat corn, which they had previously thought of as animal food. They learned what an ice cream float was, and that root beer was not beer. They would soon learn how to play horseshoes and how to square dance. They also learned that "See you later" may mean you'll never see the person again. The first time someone said this to Karl at a social gathering, he thought they were just going off into the next room and was quite surprised when they left the house. Some measure of culture shock was perhaps inevitable. They found it odd that people ask, "How are you?" when they don't really want to know that you are weak and sick from an infected shrapnel wound.

Port Allegeny is only a dozen miles south of the New York state line and the small town of Olean, where the *Olean Times Herald* was published. In summer of 1952 the newspaper ran two articles based on interviews with Karl and Hildegard. With the help of a young Swiss radio announcer who spoke German,[4] the couple discussed the forthcoming presidential election, hoping that Dwight Eisenhower, popular in Europe, would win. The newspaper also reports how the Hirschs were "amazed and delighted at the price of food and clothing." Butter cost sixty-nine cents a pound, compared with more than three dollars in Germany, and the price of German shoes was astronomical at about fifty dollars. The article went on to summarize Karl's prison camp ordeal and Hildegard's escape from the eastern sector. World War II would remain an endless source of fascination for historians and laypersons alike. Millions of people had stories to tell if they were willing, and immigrants like Karl and Hildegard added to the color and melting-pot culture for which America was famous.

The *Olean Times Herald* article was followed four days later by another article, with a subtitle referring to the Hirschs as a "Young DP Couple." The article dealt with a workshop where the subject of German education was discussed.[5] Hildegard is quoted as saying, "When children go to school in Germany it is more serious than here, fun is over." Moreover, education in Germany is compulsory through eight grades, and those not preparing for university must next undertake vocational study for two years. She went on to inform the assembly that "children are taught not to speak in school unless in answer to a question." Karl revealed that because his "father was not a member of the Nazi party," he could not continue to university without paying high tuition. From the article we also learn that since the war all children in West Berlin start learning English in second grade. Further, we learn that four hundred of the six hundred persons in the displaced persons camp where Karl worked were children, many of whom spoke Russian, German, and English. The couple also discussed the benefits of "socialized medicine" in Germany supported by a 22 percent earnings tax, half of which was paid by the employer. However, Karl expressed little enthusiasm for this system, saying he would rather save independently. He also lamented that there are three hundred thousand unemployed people in West Berlin, leading to a high incidence of juvenile delinquency. Karl and Hildegard were already telling their stories to a Western audience.

Epitomizing the aforementioned awareness and concern over the Eastern-Western Bloc split, the *Olean Times Herald* followed the 1952 articles with a 1953 piece dealing with the horrors of East Berlin under Soviet rule.[6] This time the article focused on the reports of one Miss Mary Heilner, a Union Theological Seminary graduate, who had known Karl and Hildegard as friends in the refugee camp in West Berlin. Sounding optimistically evangelical, Miss Heilner, representing the World Council of Churches, spoke of East Berlin "churches filled with spiritually awakened people who proudly wear their religious emblems and fearlessly witness their faith publicly in the face of communist domination." This, she opined, highlighted a "great cleavage between Christian and the non-Christian in Berlin" in an atmosphere "charged with tension and fear." Karl and Hildegard are not quoted or even mentioned in the article, although they and Miss Heilner beam from the photograph captioned "Reunion." Neither Karl nor Hildegard manifested the religious faith of their seminarian friend. With the Cold War and McCarthyism in full swing, neither had much appetite for politics. They had pledged to escape the very forces that could make politics so divisive and dangerous, and in any case the cynic might suspect Miss Heilner of being a bit too good to be true, and rather obviously politically motivated. Karl and Hildegard knew the plight of displaced persons (called DPs in the articles) and could tell of horrors they had witnessed, but they had seen too much to evangelize or cheerlead for any faction. They judged people by their decency, but perhaps like all who share in that mysterious phenomenon we call human nature they harbored preferences, prejudices, and resentments and the buttons that could be pressed to release those feelings under certain circumstances.

Working hard enough, sometimes twelve hours each weekday and eight hours on Saturday, ensured a reasonable income, enabling Karl and Hildegard to buy a car, their first. It was a memorable day for the DP couple, and it cheered Hildegard to no end, at the time in hospital for a small operation, when Karl drove proudly by her window. It could almost have been a car commercial. However, the couple's immigrant status held them back in small-town America, so they decided to move west again, to Cleveland. There Karl was disappointed by the petty backstabbing in the factory where he first worked, even among fellow Germans, who would do anything to get ahead. This was a subject of some import. Scientists were known to get citizenship within two to three years, whereas normal emigrants might wait five years. Karl knew of one proverbial rocket scientist, quickly nabbed after the war by the British and then by the Americans, who was given citizenship quite quickly. So, Karl decided to learn a trade and went to vocational school to become a "tool and die man."

Notes

1. Unless otherwise stated, all quotations in this chapter that are obviously in Karl Hirsch's voice come from his brief biography *Walking in Your Moccasins* (WIYM), from the Hirsch Audio Archive (HAA), or from notes made by the authors when interviewing him directly.

2. William Strauss and Neil Howe, *The Fourth Turning: An American Prophecy— What the Cycles of History Tell Us About America's Next Rendezvous with Destiny* (New York: Broadway Books, 1997).

3. http://www.navsource.org/archives/09/22/22157.htm

4. *Olean Times Herald,* June 13, 1952.

5. *Olean Times Herald,* June 17, 1952.

6. *Olean Times Herald,* February 2, 1953.

PART 3

Wild West Wanderings

CHAPTER EIGHT

The New, Ancient, and Wild West

"Free as never before," the couple loved to ski and so visited Colorado, where the mountains and clear air proved more alluring than the rivers and cities of Pennsylvania. Finally, after seven years in Cleveland, and after decades of displacement, they were Colorado-bound. Their trip to Colorado would be the final leg of their westward migration and their final chance to realize their American dream in their own permanent home. Karl remembers how he laughed at the slogan on the U-Haul truck: "Moving is adventurous." "Tell me about it," he must have thought! There are travel adventures and there are travel adventures. Being stuck in an airport for a few hours or even overnight is not worth the telling. Walking from Romania to Stalingrad to get a buttockful of Soviet shrapnel is another story. Marco Polo and Ernest Shackleton had travel adventures, long walks, and narrow and great escapes. Likewise, we know Karl's experiences were worth telling, because people asked to hear them but, moreover, because they shed fascinating historical and sociopolitical light on global upheaval and the turbulent displacement of people in the so-called civilized world.

In retrospect, these stories were not simply about surviving prison camp in the Eastern Bloc and escaping to the West. All manner of political currents swirled in and around the German, Polish, and Russian/Soviet expatriate communities and the precarious cultural status of immigrants. The Germans had been enemies of the Allies, but the Russians, although they had been allies who had helped defeat Germany, were now Communist "bad guys" in the eyes of many Americans. But what was one to make of the German scientist Karl met in Cleveland who was working for Bell Aircraft, no less, yet had a

Although Joseph McCarthy had served as a marine in the Pacific theater, rising from the rank of lieutenant to major, he clearly had little inclination to see the Soviets as allies. Unlike Karl, who shied away from the officer and political classes for ethically driven reasons of conscience, McCarthy had political ambitions and would be unmasked for lying about his supposedly heroic military exploits, which were exaggerated, indeed fantastical. Such unethical maneuvers did not prevent his postwar career in politics, but it was short-lived. It is for his use of unfair allegations and investigations, which came to be called McCarthyism, rather than any humanitarian deeds that he is remembered. By the time Karl and Hildegard settled in Denver, McCarthy had been censured by his fellow senators, one of whom accused him of being "the kid who came to the party and peed in the lemonade." He would be a sad and lonely figure when he died, at the premature age of forty-eight, a victim of hepatitis and alcoholism.[1]

picture of Hitler on his living room wall for all to see? Did this prove him a Nazi and absolve him of any potential Communist sympathies? Surely one would suspect white supremacist prejudices? The infamous senator Joseph McCarthy, a fervent anticommunist who had been elected to office in 1947, in the year of Karl's release from Soviet hands, would no doubt have approved of, or at least turned a blind eye to, Germans with anti-Soviet sentiments. Scientists and engineers, especially the rocket science breed, had often been in demand in the military-industrial-weapons industry and in wartime could face pressure and threats from opposing sides. It is perhaps for such arm-twisting pressure that ostensible enemies (Germans) would make their side-switching allegiances clear by espousing anticommunist ideologies. Such issues of allegiance, clouded by suspicion of such persons by both Germans and Soviets as well as among Americans, would be a factor pertinent to Karl's future as he began to use his newly acquired tool and die maker training to apply for jobs as a skilled machinist.

The Rocky Flats Plant, a nuclear weapons production facility, had been in operation for a decade before Karl secured a job there in the early 1960s. It took him two years working at other machinist jobs to obtain clearance from the FBI. They were thorough in their acquisition of background information, yet not surprisingly were unable to account for his whereabouts during his two and a half years in Siberia. They asked him questions such as "What would you do if 'they' (the Communist bad guys) put pressure on your sister back in Berlin?" We do not know Karl's explicit answer, but he recounted an incident in West Berlin when Karl and his boss caught known thieves stealing two hundred yards of telephone cable. The thieves were taken to court, where they openly threatened Karl from the stand. But with no leg to stand on, they were denied asylum and deported. The record of this obscure judicial proceeding evidently figured in the background check as evidence of Karl's solid opposition to crime, delinquency, and lack of sympathy for the Eastern Bloc regime. Karl eventually saw a copy of his FBI file, after retirement, and was gratified, though perhaps not surprised, to learn that he and Hildegard had been characterized as good people who "kept themselves to themselves."[2] They had always been sincere, since the late 1940s, in their joint pledge to build a simple, decent, honest, and better life, away from the old country and the tired Old World.

Karl and Hildegard bought a house in the Denver suburb of Lakewood and tried unsuccessfully to persuade her sister, Margaret, who had lost her husband during the war, to let them adopt her son Sebastian, the oldest of her five children. In the aftermath of refugee camp experience, perhaps this would not have seemed an unusual arrangement, especially for a couple who might be childless for any number of reasons. In recent decades Karl and Hildegard had rarely experienced traditional nuclear family domesticity, and it still was not to be. The impossibility of adopting Sebastian was a factor in the childless couple deciding to adopt two boys, brothers already aged eight and nine. Karl would admit that they turned out to be "more than we bargained for." They had, Karl would write, "already been spoiled," by which he did not mean pampered so much as ruined by having suffered an atrocious upbringing. It was reported their mother had been a prostitute, and evidently there was no father in the picture. In short, both boys would prove criminally incorrigible. In fact, one had been involved in the murder of a sibling when only seven years old and, one would hope, too young to understand the enormity of such a crime. Again, perhaps in the aftermath of his refugee camp experience, Karl thought he and Hildegard might whip the boys into shape. They would not be the first juvenile delinquents the couple had had to deal

Karl and Hildegard smile as they show the fossil eggs that launched Karl into a new field of paleontological research. Photo from 1977 in the University of Colorado Museum archives.

with. However, neither charitable good intentions nor the stern discipline Karl had sometimes been forced to impose in the Berlin youth house and refugee camp worked in Colorado, at least not with any sustained success. Their hope of integrating these boys into their family would ultimately prove to be in vain.

Karl would not mince words in saying that despite trying to polish the boys up, "they were still rotten inside" and he and Hildegard "did not quite know how to handle them." Both boys ended up doing serious jail time as young men. Karl would later see the bizarre, even funny, side of his and Hildegard's unlucky adoptive parenting experiences. He would recount sitting across from an attractive and sophisticated conference delegate who proudly extolled the achievements of her successful children who had breezed through Ivy League schools to coveted professional careers. When she asked him about his kids, he could only laugh rather nervously, already knowing he would astonish her when he informed her that both his sons were in jail!

Naturally, perhaps, Karl had hoped to adopt a girl, but Hildegard said she preferred boys, who did not "bring their problems home." Sadly, she was proved wrong. The boys had had a poor start before coming into the Hirsch home. Neither could really read or write properly, tell the time on a clock, or, and this stuck in Karl's memory, distinguish between vowels and consonants. However, they did have the chance to learn Hirsch-household German, and the older boy would spend two years in the army in Germany. The younger boy was habitually in trouble. He stole from the local McDonald's restaurant, stole from Karl, and even stole from guests, which Karl considered the most egregious of sins. This led to big confrontations, including visits from the police and from probation officers, one of them, as Karl would remember vividly, a young woman in hot pants. In the circumstances, this sign of the casual times appeared disconcerting to an older, conventional, German couple trying to impose a serious sense of socially acceptable conformity. As a seasoned con man, the younger boy managed to play on the probation officer's liberal sympathies by buying her a hamburger with money stolen from the household! On an occasion when he was arrested in Lincoln, Nebraska, he even conned the local sheriff and escaped across state lines back to Colorado. Eventually, as a remedial last resort, he joined the marines and made it through boot camp. But this did not lead to a long-term rehabilitation. He would again fall by the wayside and into the prison system.

Karl and Hildegard took to the outdoor life, and they would sometimes take the boys camping with them in the hope of toughening them up in the bracing mold of healthy, rosy-cheeked, cold-water-bathing woodland scouts. Karl admits that once, after being forced to lug campfire water himself because the boys refused the chore and were even too sullen to reply to a well-intentioned "Good morning" greeting, he dumped the whole bucket over one of the recalcitrant heads. It did not bring about the hoped-for transformative remedy. It is interesting to ponder what the boy's response might have been to the authoritarian question asked of Arlo Guthrie, and retold in his famous 1967 rebel song "Alice's Restaurant": "Kid, have you rehabilitated yourself?"[3]

Karl and Hildegard rehabilitated their own battered psyches with the tonic of crisp, ski-slope air and hands-on, earthy grounding and grubbing with rock-club friends, cobiographer Bernard (they called him Bernie) among them. Banging with rock hammers also must have released pent-up frustrations. Their first field trip was to an area north of Moab, Utah, a good five-hour drive from Denver even on today's freeways, where they found dinosaur vertebrae and fossil wood. Karl had intuitive, conservation-conscious misgivings about removing souvenir bones from articulated skeletal remains and thus compromising the integrity of the whole fossil assemblage. In fact, excavation of vertebrates was illegal without a permit, but in those days awareness of the rules was as lax as their enforcement. There was a tacit awareness that a species of desert rat one might classify as the "bone bandit"—related to the pot hunter, scourge of the archeological world—operated subversively in the dusty desert shadows. The club rockhound was another, more responsible and semiprofessional species, but the role of club leader and tagalong professional in discriminating between legal and illegal collecting was still somewhat murky. What we do know, then as now, was that much of western Colorado and Utah is public land, where camping and campfire wood are free, as are the mental health benefits of escape from the city and from workplace drudgery. The fossils, except those of vertebrates, are also free to collect, if not to sell legally, although rules and guidelines, as well as enforcement strategies, have changed over the years.

Karl and Hildegard had come to an outdoor paradise, a mecca for hikers and for amateur and professional gem, mineral, and fossil collectors. Martin, cobiographer with Bernie, recalls a geology student from the Midwest reaching the mountains, grasping a rock in each hand, and whooping, "I can't believe it. I'm in the Rocky Mountains." Colorado is the only state with a town called Dinosaur and another named Bedrock. There are also towns, settlements, and locations named Basalt, Castle Rock, Coal Creek, Gypsum, Hot Sulphur Springs, Leadville, Marble, Monument, Palisade, Rocky Ford, Silt, and Silver Plume. The state is a veritable geological paradise and, as we shall see, has been fertile ground for the fossil hunter since Colorado achieved statehood in 1876, just a year before a major dinosaur discovery at what is now Dinosaur Ridge, west of Denver a few miles from where Karl and Hildegard lived.[4] Although he did not know the trivial fact at the time, Karl's paleontological explorations were beginning almost exactly a century after the first discovery of a fossil egg, in England in 1859, the same year that Charles Darwin had published *On The Origin of Species*.

Trips to marine deposits in eastern Colorado, Kansas, and Wyoming would spark interest in ammonites, nautilus-like fossil favorites and members of a molluscan clan much studied by the professionals. These same fossil fields, among the first excavated in the West in the 1860s, would also yield the skeletal remains of giant marine reptiles. Campsite camaraderie and fossil fascination also had the reward of dissolving any amateur-professional distinctions. Geologists, after all, are earthy people, and rock-club enthusiasts like Karl and Hildegard could mingle with US Geological Survey experts like Bill Cobban, famous for his ammonite expertise.[5] All had equal chances to make significant finds, in line with the club's collective and amateur science goals, and there was also a generally responsible "We love and respect fossils" ethic.

Karl and Hildegard were able to bring order into the more chaotic aspects of their new life by running the household in an orderly way, with commonsense discipline over domestic affairs and workday routines. Weekdays for work and weekends for healthy exploration. Finances were kept in order despite the boys' transgressions, the dogs were well behaved, and the fossils were classified and put in cabinets with the same reverence that some people bestow on heirloom china or sports trophies. Fossils were cleaned up as necessary and given their appropriate name tags and correct scientific names. They may not have said "Thank you" or a polite "Good morning," but they were silently obedient and did not talk back. Even if the occasional specimen was under suspicion of having come from a stolen-goods neighborhood, fossils kept a prudent silence and behaved well. Karl and Hildegard could be comfortable with, even proud of, their fossils in ways that they could not be with their difficult, guest-robbing boys. Their fossils even had respectable pedigrees, having been obtained ethically on organized expeditions and not by shady bone bandits or pot hunters.

Such orderly fossil classification methods had originated back in the 1730s when a distinguished Swedish naturalist, Carl Linnaeus, published his famous *Systema Naturae*, for the first time giving formal scientific names to plants and animals.[6] This classification method was universally adopted, and used for fossil species as well as living species, under the name of Systematics or Taxonomy, the latter literally meaning "the arrangement of names." Little did Karl Hirsch know, after putting name tags on weekend finds, that he would soon be adding new systematic vocabulary and new taxonomy to paleontological glossaries and monographs.

If TV and media are a reflection of society, the popular courtroom drama *Perry Mason* gives insight into American culture in the late 1950s and early 1960s, when the episodes were filmed. Interestingly, almost every episode involved a defendant falsely accused by overconfident authorities and then vindicated by clever detective work by a brilliant attorney. Behind the drama one notices that despite their mostly formal and conventional dress and generally polite behavior, both the good guys and gals and the bad guys and gals go in for a lot of smoking and drinking. Many plots revolved around actors with alcohol problems, while drunk driving seemed to be commonplace. Was this just on TV or was it a reflection of real life? Historians and psychologists tell us the one always reflects the other.

Karl and Hildegard were driving back from a rockhounding trip to Utah, in their large Jeep Wagoneer. Hildegard was cleaning a bag of fresh Western Slope peaches in a bucket of water while Karl drove. It was late in the summer of 1961 as they descended the last few miles of the Front Range Eastern Slope back into Denver on a perfectly good three-lane highway. It had been a good trip, as proved by the haul of dinosaur bones and peaches. A drunk driver veered across the road, hitting the Wagoneer near the rear driver's side wheel, turning the whole vehicle ninety degrees hard left. The Jeep bounced and rolled over several times, ending right side up as Karl was knocked out by a hefty chunk of airborne dinosaur bone. Such are the perils of paleontology.

Photograph of the wrecked Jeep Wagoneer after the roll-over accident caused by a drunk driver.

Fortunately, both Karl and Hildegard were wearing seat belts, because although the sturdy Jeep only had its roof flattened by about a foot, everything inside was broken or damaged. Karl later noted that the gas can had suffered sharp dents but thankfully had not been pierced. He would never again carry a gas can inside his vehicle. As emergency vehicles came and took the couple to a local hospital, a huge traffic jam ensued. Other returning members of the rock club recognized the battered vehicle and came to rescue the scattered dinosaur bones and other belongings and to inquire after Karl and Hildegard. They were told by the state trooper that he could divulge information only to relatives. The club members pleaded that the Hirsch couple were displaced persons and had no relatives west of Berlin, but the trooper was tight-lipped, forcing the group to call every hospital in the area. Annoyed by the trooper's lack of cooperation the club members filed a complaint, which landed the trooper a reprimand.

Karl tried to go back to work as soon as possible, but Rocky Flats proposed putting him on sick leave at the reduced pay of $40 a week. Karl said he could not afford such terms and would quit if he could not quickly resume work full-time. Naturally enough, Karl hoped for compensation from the guilty party and his insurance company. The other vehicle had been occupied by two construction workers who had provided ample evidence of their drinking habits in the litter of beer cans in their vehicle. Despite a DUI fine, ironically of only about $40, the driver had no insurance, so no settlement was forthcoming and Karl had to go to court. There he was told off for appearing in shorts, which he was wearing because, as a wounded victim, his leg was still bandaged. It was a year before Karl's insurance company attorney, evidently not of Perry Mason caliber, was able to secure a settlement with the guilty party. The delay was in part because one of the culprits, and a potential witness, had since died in another vehicular accident, this time involving a bulldozer! Drunk or sober, we don't know, but evidently some people court trouble. The living defendant's lawyer eventually proposed a final offer settlement of $500, backed up by the threat that his client's alternative was to declare bankruptcy. In the end, Karl's insurance company received $400 and passed $100 of it on to Karl, which he spent on a microscope. Such, indeed, are the perils and limited financial rewards of paleontology. But it is perhaps a rare travel story in which the protagonist is knocked out by a flying dinosaur bone.

Notes

1. R. Cohn, *McCarthy* (New York: New American Library, 1968). Famously, Senator William Jenner told Senator Joe McCarthy, "Joe, you're the kid who came to the party and pee'd in the lemonade." In other words, McCarthy wouldn't shut up and just go along to get along.

2. A letter to Karl from the United States Atomic Energy Commission (Rocky Flats Office) dated July 11, 1962, refers to Karl's "security clearance," inviting him to come in for an informal interview.

3. "Alice's Restaurant" is Arlo Gutherie's most famous, serious, yet tongue-in-cheek Vietnam Era antiwar song. It represents the antiauthoritarian sentiment of 1960s youth and serves as context to the challenges faced by Karl and Hildegard in

trying to instill values in their two already recalcitrant adopted boys. Arlo Gutherie was son of the famous social-reforming folk singer Woody Gutherie (1912–1967), who championed those who were downtrodden by authority, a position not unfamiliar to Karl in his soldier's view of the privileged officer class.

4. The destination known today as Dinosaur Ridge is an historic paleontological site, where Jurassic dinosaur bones were discovered in 1877 and Cretaceous tracks were found in the 1930s. The Jurassic bones come from the world-famous Morrison Formation, from which Karl would find and study dinosaur eggs. Karl lived only a few miles away from Morrison and lived to see the Dinosaur Ridge nonprofit be founded in 1989 and enjoy healthy growth in his lifetime.

5. The following is taken from Bill Cobban's obituary (published in the Geological Society of America *Memorials*, vol. 44, May 2015): "Dr. W. A. 'Bill' Cobban, one of the most highly respected, honored and published geologist-paleontologists of all time, passed away peacefully in his sleep in the morning of 21 April 2015 at the age of ninety-eight in Lakewood, Colorado. Bill was an extraordinary field collector, geologist, stratigrapher, biostratigrapher, paleontologist, and mapmaker who spent nearly his entire life working for the U.S. Geological Survey (USGS)." See https://www .geosociety.org/documents/gsa/memorials/v44/Cobban-WA.pdf. Cobban accumulated numerous professional awards and accolades. He was also remembered, however, for how he "shared his knowledge with all those around him and with everyone who visited him in Denver or accompanied him in the field."

6. The Swedish botanist-scientist Carl Linnaeus (1707–1778), also known as Carl von Linné, is famous for his great work *Systema Naturae* (1735), which is known to biologists and paleontologists as the basis for the formal scientific naming of species using the binomial (two names) system, as in *Homo sapiens* or *Tyrannosaurus rex*.

CHAPTER NINE

Do-It-Yourself Paleontology

After getting into the swing of Colorado fossil-hunting culture in the early 1960s, and rubbing shoulders with professionals like Bill Cobban—who encouraged systematic collecting of ammonites, an easy choice given his specialty—Karl and Hildegard became steadily more enamored by fossils and paleontology. Social and field excursion opportunities afforded by the rock club had offered an entrée into the wider paleontological and geological community that thrived in Colorado, an oil, gas, coal, and mining frontier and base for many active companies. Denver has many colleges and universities, and its museums swarmed with geologists of many stripes. It is one of the few cities with a large US Geological Survey complex, the official geological arm of the federal government. Long drives and campfire nights in the company of motley crews of enthusiastic and knowledgeable professionals and amateurs were as informative and educational as the few seminars and classes Karl would take to integrate his growing knowledge. Members of the group knew where to collect in half a dozen western states, the best access routes once you left the main highways, the simple tools you might need, and whose permission you needed when accessing private land and even certain public land locations. Members also had experience in spotting half-buried fossils and in extracting and transporting them most efficiently without damaging them.

Some of the best ammonites, for example, came from fossil-rich marine deposits known as the Pierre Shale, a gray shale formation laid down when a shallow sea flooded the interior of the region that is now the US during the latter part of the Cretaceous period, some seventy to eighty million years ago. Such ammonite treasures, along with complete fish and large marine reptiles known as mosasaurs, could be found in the Red Bird region of Wyoming close to the borders with Nebraska and South Dakota. Like Moab, Red Bird was about three hundred miles from Denver, an opportunity for a good five or six hours of paleontological gossip each way. Ken Carpenter, whom we shall meet again soon, recalls first meeting Karl in the field at Red Bird in the mid-1970s, when he rendezvoused with the University of Colorado vertebrate paleontology class run by then-professor Judy van Couvering. Judith, as she preferred to be called, was a positive influence on Karl and lovingly mentored him through his scientific journey, which sometimes involved lively, stormy scenes.[1] The University of Colorado Museum of Natural History had a special interest in the Red Bird area, having recently been given a large collection of vertebrate bones (of both fish and marine reptiles). From ammonites to fish and reptiles, the paleontologically inclined enthusiast could clamber about the prehistoric tree of life from buried branch to petrified limb.

If you preferred a shorter excursion, say, near Pueblo, only one hundred miles south of Denver, another type of ammonite was famously abundant at Baculite Mesa, a fine example of a place named after a fossil—that is, the ammonite *Baculites*. In such places, and there are many, if you avoided driving on the shale when it was wet, axle-sucking gumbo, you could find gullies that had not been picked over and you could net a bucketload of fine specimens. It was not only the terrain and weather that created obstacles to fossil fields. One of the sites near Pueblo that Karl wanted to collect from was owned by the Arizona Cattle Company and had the hillbilly-sounding name of Tom Hollow. Bernie remembers that Karl needed his help to get access, which Bernie could facilitate by playing his professorial University of Denver card when meeting an owner-representative in Denver. Phone calls to the old-time resident rancher ensued, followed a few days later by a face-to-face meeting with him, his pipe-smoking wife, and two dangerous-looking dingo dogs. Karl and Bernie were regaled with inevitable tales of how the dogs were used to round up trespassers, and neither would soon forget the pipe-smoking gal who reminded them of the character Mammy Yokum from the *Li'l Abner* comic strip, who resembled a female Popeye the Sailor Man. Once safely through the human and canine chain of command, the fossil hunters had access to Tom Hollow's ammonites, and Karl would tell Bernie that the hollow's gullies reminded him of Stalingrad "with Russians, just on the other side."[2]

Ammonites and other fossils still encased in shale could be taken home, and "prepared" for the display case by removing all the shaley matrix that had entombed and protected them for the last seventy to eighty million years. Properly prepared, many of these ammonites literally reveal their true colors, iridescent mother-of-pearl shells that are among the world's finest, exhibit-quality treasures. Although illegal to sell if from public land, those from private land, and strays from who knows where, fetch steep prices in high-end rock shops the world over. The "gold in them thar' hills" out west may be the dazzling mother-of-pearl sheen of extinct mollusks.

There is, or was, more so in the past, a murky intersection between the amateur, professional, museum, and commercial worlds of paleontology. For example, one cannot legally sell fossils from public land or even collect vertebrates without a permit. However, museums can pay preparators to clean or "prepare" specimens, and such tasks may take weeks or months, not to mention special skills and patience. Karl, whose interface with fossils was always honestly aboveboard, would amass a fine and valuable ammonite collection, which he would eventually donate to the Denver Museum of Nature and Science. Meanwhile, amber caught Hildegard's eye, and she began collecting specimens that were available on the open market, some coming from the Baltic shores of her native Germany, washed up on the coast where Karl had bathed as a teenager, and where he had later been captured by the Soviets. Hildegard's specialization of choice, from among the myriad insects and other "bugs" trapped in this gem-producing fossilized resin from forty-five-million-year-old pine trees, was the spiders, one of which had been given to her by David Armstrong, a friendly University of Colorado professor.[3]

Many classic and interesting chapters in the modern history of paleontology have been written by amateurs, and in truth there were very few specialist professionals in the field before the late nineteenth century. Many of the most celebrated names from that time were natural-history polymaths, studying everything from butterflies to barnacles. However, by the latter half of the twentieth century professional paleontology had become considerably more specialized, as more and more discoveries were made. Many graduate students would specialize in the study of a particular species or group of species from a relatively localized area or a narrow slice of geological time. If entering the ranks of university or museum professionals, they might continue such specialized study for an entire career. These were marked men and women:

"He studies ammonites." "She studies mammals from Wyoming." "Him? He spent years in that *Allosaurus* quarry in Utah." While club amateurs were free to roam widely, professionals too often tended to cloister themselves in self-imposed prisons of specialization, what cynics might misleadingly call their proverbial ivory towers. All pursuits have their occupational hazards. While a few irritable amateurs have complained in print that their finds have not been recognized or that they cannot even get the professionals out in the field to bless "their" favorite discovery sites, most realize that professionals are friendly allies who are stretched thin and that they depend on them to help with research, publication, curation, advice, referrals to other specialists, and so on.

In the 1960s the Wild West was, and still remains, a regional and still-wild patchwork of expansive wide-open spaces where there are tensions between private landowners and those who use and manage public land, especially where boundaries are poorly marked, or where private owners have grazing rights on public land. There are legendary stories of ranchers painting their cattle orange in hunting season, so the animals would not be mistaken for deer or elk. It is even said that some ranchers have written "cow" and "horse" on the appropriately labeled species, almost as a fossil hound would correctly label an ammonite.

Karl would chuckle over memories of hunting season encounters. Once in antelope season Karl, who also sometimes hunted the species, ran across a rancher marching a hunter off to the sheriff for hunting among his herd of cattle. On another occasion hunters, who had bagged an antelope, asked the rockhound group what they had bagged. The chuckled reply was "two fossil fish and a mosasaur."

From Colorado to Utah and Wyoming, Karl was priming the pump that would propel him into ever more professional, paleontological endeavors. Famous as these states are for their age-of-dinosaurs fossils, from ammonites and fish to marine reptiles and dinosaurs, the region is also famous for younger fossils from the age of mammals and birds. On behalf of the University of Colorado Museum of Natural History, Karl, Hildegard, and their club compadres would often collect mammal fossils from the Nebraska badlands, where it was relatively easy to collect complete skulls of twenty- to thirty-million-year-old sheep ancestors, known as oreodonts, and where sites had been found yielding literally thousands of small rodent jaws.

Paleontologists jokingly describe a certain species of mammal paleontologist as paleo-dentists. The University of Colorado paleontologists were at the time much focused on mammal fossils from Wyoming and Nebraska. Karl played a significant supporting role by collecting for the university museum for almost a decade. This gave him a chance to brush up his paleontological credentials by taking a couple of courses in vertebrate paleontology from the aforementioned professor Judy van Couvering, whom we shall soon introduce more fully by her maiden name, Judith Harris.[4]

It was in these jaw-strewn Nebraska badlands that one day, in 1973, the fifty-two-year-old rockhound "looked down and there was a perfect egg," which he at first thought was "a very old duck egg." In retrospect, this gift was career-changing for Karl. It had come from a bird that had laid a symbolic golden egg in his path. Not shy about acknowledging the mysterious forces of destiny, Karl would say the egg was "put there for me"; a blind man could have seen it. This finding would launch Karl into two decades of study in the specialized field of egg and eggshell research and propel him to the status of one of the elite experts in his field.

It would be six years between this, his first egg find in Nebraska in 1973, and the publication of his first paper on fossil eggs. There are no rules for transitioning from rockhound to academic specialist, from amateur to professional, and in any event these labels are rather artificial. Some so-called amateurs, exercising huge, even obsessive love and dedication to paleontology, have amassed impressive experience, knowledge, and publication records and are amateurs only because they have not had formal institutional appointments or paid careers in paleontology.

Karl would do what was needed to professionalize himself with a combination of common sense, trial and error, individual determination, and help from friends and colleagues. It was do-it-yourself paleontology, with "a little help from his friends." The obvious first step was to show the find to a knowledgeable expert in the field. It is now part of Karl legend that he could find no such expert, dare one say "eggspert." However, birds lay eggs, so it made sense to consult an ornithologist, preferably one who knows fossil birds. Karl was able to consult one Alexander Wetmore, already an octogenarian who had long ago worked in Colorado, Kansas, and Idaho and was the author of *A Checklist of the Fossil and Prehistoric Birds of North America and the West Indies*, a book now vanishingly rare.[5] Wetmore, however, was little help and could only tell Karl the unvarnished truth: that there was no one studying fossil eggs in North America and precious few anywhere. One could in fact count those interested in the subject on the fingers of one hand. The ink was barely dry on three papers published in Russian by Andrey Sochava between 1969 and 1972, and they had likely not been read by anyone west of the Iron Curtain, and perhaps few in the Eastern Bloc.[6] During this same period (1970–1972) a German paleontologist, Heinrich Erben, would publish three papers on fossil bird and reptile eggshell.[7] Erben would be the first of Karl's few early contacts, a colleague who literally spoke the same language, and with an interest in this specialized field. As Karl often told his friends and colleagues, he had basically been informed that if he wanted to learn anything about fossil eggs he would have to "figure it out" for himself. Undaunted by the challenge, this is exactly what he did.

However, Karl had many supportive colleagues in Colorado, and he also took the initiative to write to those, like Erben, who had published on the subject. Erben was also an early pioneer in the use of the electron microscope in what he would call not microstructure, but ultrastructure, studies. He had, Karl would learn, a good scientific team, including colleagues who worked on ostrich eggshell, which is scattered abundantly around the younger archaeological sites and older fossil sites of the Old World.

You do not necessarily have to publish a scientific paper to be labeled "one of the foremost experts in the world, researching and categorizing fossilized eggs." This description appeared in 1977, in *Rockwell International News*, or *Rockwell News*, the magazine associated with the Rocky Flats nuclear facility where Karl worked. Clearly, the company was proud of its creative employee, before his first paper was ever published. In fact, they would help sponsor his research by matching his out-of-pocket expenses dollar for dollar. Legend has it that Karl cleverly "wangled" this support by representing himself as a nonprofit entity: thus, each dollar he gave himself was effectively doubled. Clearly, Karl's experience in the supply requisition and quartermastering business had not been in vain!

By the mid-1970s, Karl was already known among his paleontological peers for his interesting, specialized work. Although it does not show up on his published bibliography, according to *Rockwell News* he had presented his findings under the title "Fossil Eggs of the White River Formation (Oligocene) of North America." This was, again, a landmark for a man his employers described as an "ex-German civil servant . . . with very little formal education." Bravo! His mentors, the professors van Couvering (both Judy and John), reported that the presentation was received by many in the audience as "one of the best." Stressing the newness of the field, Karl stated, "There are maybe two people who study . . . eggs in Germany, maybe one in Russia and a few in France. But I think I am the only one in the United States, except for James Jensen."[8]

This latter reference was to the famous "Dinosaur Jim Jensen," who had a lot in common with Karl who was only three years his junior. Having had little formal education, Dino Jim tried homesteading in Alaska but was "displaced" by the construction of military installations during the war and so moved to Utah where he took a crash course as a machinist and even worked on a nuclear reactor. A competent artist and sculptor, he then veered into museum exhibit and fossil preparation, spending a lot of time in the field working closely with rockhounds. He worked tirelessly to help establish the Brigham Young University paleontology program, impressing professional colleagues with his energy and his authentic curiosity for things pertaining to the natural world. As two of his younger colleagues wrote, "Much of his success stemmed from the local contacts he made with people familiar with the Utah backcountry, and his charismatic public persona."[9] He became most famous for huge dinosaur excavations in western Colorado, and for amassing equally large dinosaur collections composed of monsters with suitably gigantic names, such as *Supersaurus,* which went viral around the global media circuit in the days when "viral" meant glossy magazine and TV documentary coverage. Brigham Young University gave Jim an honorary doctorate for his services to the field. Karl would get one too, but not until 1990.

Karl recalls that Jim was "moderately helpful at first" but then proved reluctant to share his eggshell finds. Karl was undeterred and got his own permit so that he could collect by himself. Jim reportedly complained that some claim jumper had invaded his turf. They later met when Karl was visiting the site with other paleontologists, including friend Judith, and despite gruff grunts at first, Jim warmed up. *Rockwell News* ended its 1977 article, prescient at the time, with the prediction that "Karl Hirsch is probably destined to become the top expert on fossilized eggs in the United States, maybe in the world."

The glowing words penned by *Rockwell News* about Karl's fast track to scientific fame could not completely disguise the inevitable frustrations he faced in writing his early papers in technical, scientific English. He relied heavily on Judy for help in these endeavors. Throughout the 1970s they would meet in the field, the classroom, and the museum. Some witnesses remember their squabbles over manuscript preparation as "testy" and "occasionally very loud," but their relationship was one of abiding friendship and mutual respect. They shared many friends and colleagues in the paleontological community, including we cobiographers. While Karl deserves all the credit that he has been given for successfully mastering difficult analytical techniques, amassing an impressive collection, and making a lasting mark as a paleontologist, Judy, who later reverted to her maiden name Judith Harris, was an important, warm, and friendly breath of wind beneath Karl's wings at a time when he needed moral as well as editorial support. At the risk of psychological digression, we believe she perhaps gave him insight into the role women of her generation were playing in a science that until then had mainly been the province of men. It was a good, if subtle, pointer to his future, in which a number of younger women would come into the field and collaborate with him. They may have been substitutes for the daughters he never had, but some, like his friend and scientific colleague Emily Bray, would recall that he sought the comfort of female companionship, so different from the officious, rank-conscious stance of the soldiers he had sparred with in the German military. But these friends were also indicators of the changing times, signs of a sea change in women's involvement in science. Women were knocking at the doors of university departments to say, "If you're doing science here, please include us. We have much to offer." Today there are about as many women as men in the professional ranks of paleontologists.

When Karl was prompted by the Harvard paleontologist Farish Jenkins to study what was reported to be the oldest fossil egg, Judith told him, "You can do it . . . you are careful and thorough enough." This would prove to be the case (part 4). Karl did not manifest any obvious lack of confidence, but he was working full-time and would admit that he was not the supremely driven type. He saw Hildegard as the competitive one in the family. So perhaps, as he admitted, Judith inspired him a lot; it was something he needed, especially when his health faltered. He had been beaten up by almost a decade of war and another decade of austere recuperation and grubby labor. He had not had time in the trenches or at his blue-collar factory job to read the classics of science and literature or to discuss the finer points of existence beyond life, death, and "nature red in tooth and claw." Now, however, he was rubbing shoulders with PhD paleontologists who could ruminate on the immensity of geological time, the mysteries of evolution, and the finer points of the architecture of enamel in the teeth of extinct rodents.

The immensity of geological time is one thing. The mortal span quite another. Only a year after he published his first major paper in 1979,[10] Karl underwent a serious, confidence-sapping heart operation. Again, along with Hildegard, Judith was there to offer support, telling him, "What good is all this work if you don't pass it on." Again, she was offering advice that would prove prescient, and Karl would rise to the challenge, shaking off the shadow of ill health. There were others in the eggshell field, like Mary and Gary Packard, a husband-and-wife team, zoologists from Colorado State University who studied modern bird and reptile eggs. Mary, Karl would recall, also taught him a lot, as *Rockwell News* succinctly put it: he was gaining "knowledge in Chemistry, Biology, X-rays, Crystology (*sic*), and many other fields."[11] [The field is Crystallography.]

It is a strange paradox that PhDs from Harvard and elsewhere would essentially tell him, "If you want to study fossil eggs you will have to go it alone," while his mentor Judith and his wife Hildegard offered the encouragement that told him he was not alone. They had faith in his endeavors even if his health reminded him of his mortality. The scientific field of study that was fossil eggs might be small and at times lonely, but it was offering him a new beginning. It was surely a metaphor for the new start that life had offered him in the Wild West and the new life chapter offered him after he survived his heart operation. And what better place to start than with an evolutionary beginning, with Harvard's specimen and his own curiosity about what purported to be the oldest known fossil egg? Karl began to put his thoughts and observations on paper, and with a little help from his friends, shepherded his first few articles to the desks of journal editors. By the end of 1983 he had a half dozen articles to his name, a couple of them sole-authored and mostly his own work, a few others penned with the help of collaborators. He had not quite gone it alone into the field. It had been, and would remain, a collaborative effort.

In another cruel twist of fate, Hildegard Hirsch (HH), who had been born on Christmas Eve in 1924, lost her battle with cancer in the winter of 1984. She was not yet sixty years old. The impact on Karl can only be imagined and told in his own words and in his own time. A year later, on the anniversary of her passing, he would write this:

> It was a hard year. There were times when I wished I would have died too.... However, there were times when I had hope, when I wanted to live ... this year ... was a constant fight with myself ... I thought I should be able to handle the situation. When HH realized that she could not beat the cancer ... we talked a lot about almost everything ... "Keep working on your eggshells" [she said] "move to Boulder and marry one of the girls, so someone will take care of you." Easier said than done. I still live in our, that is my house.... Yesterday was my first day of retirement. Another big cut in my life that will add to the loneliness....
>
> The other day I talked to David [a friend] ... without sobbing ... some improvement. Told him there were mornings when I do not want to get up ... wondering why I had to wake up.... There are times when a car approaches in the wrong lane ... and you ask yourself "why not." It is scary, I do not want to hurt other people.... David listened. It is easy to tell him things.... He told me ... I have to make up my mind.... HH wanted me to go on, I hold that in front of my eyes, and the heart quits aching. I would have expected HH to go on, told her so when I had my open-heart surgery ... then I had a strong will to live ... felt ... I would make out OK....
>
> I still have a lot to do, and I do not want to let my friends down ... quite a few ... want me to go on ... most of them really mean it. Being together with them is now the sunshine of my life. I go home and feel the loneliness double heavy. That's when I talk to HH, ask to help me overcome these dark moments ... when I sit down and cry, I never thought I could cry. Cry for HH, no I think I cry more for myself, that I am alone, have nobody to share with, to share the good and the bad, that I have nobody that worries about me, wants me home, wants to talk to me, loves me....
>
> Friends are OK. They think of me as a nice person, want to have me around.... In their way they probably love me. The younger ones think what a nice old man, has seen a lot in his life, he can tell quite some stories....
>
> The work on my eggshells. Still have to write a whole bunch of papers and find someone to carry on my work, so it does not only collect dust.... I like to do it, it keeps me going ... often I think I made too many commitments [but] I never backed out of things....
>
> Often I feel like getting drunk, but after a few cognacs or so it's no fun, no fun to get drunk alone.[12]

By the time Karl penned this anniversary letter, he had turned sixty-four years old. The Beatles had asked, "Will you still need me, will you still feed me when I'm sixty-four?" The number had become iconic for retirement age, a change in life's schedule and pace when loneliness might well intrude. If the "Will you still" question had been asked to Karl directly, he would have confessed to the loneliness so sadly poignant in his letter to HH, and he might have revealed the heartfelt sadness that war and the other outrageous slings and arrows had imposed on his destiny. But the human spirit is often more resilient than the more outrageous details of biography might suggest. The slings and wounds may build strength and character. With the help of friends who he loved and who "probably" loved him, and saw his strength, and shared the occasional heartfelt tear, Karl kept on going. His eggshells, his friends, and the work he liked to do, the work that kept him going, would sustain him and be the sunshine in his life. The Colorado sun shone steadily, year in, year out, for an average of 310 days a year and would continue to shine on Karl and illuminate his life, his field excursions, his walks with his ever-faithful dog Maggie, and the work he would continue for the decade to come.

Notes

1. Judith Harris (1938–2019) was a Professor of Paleontology at the University of Colorado, Boulder. Hailing from Oklahoma, Judith received her PhD from Cambridge University in England before she and her anthropologist husband John van Couvering took curatorial positions at the University of Colorado Museum. She was initially known professionally by her married name; later she reverted to her maiden name. She made her mark publishing on African paleontology and mentoring students, including latecomers like Karl. In later life she became involved in the Women's Studies field as it became part of the university curriculum.

2. Here we continue using direct quotes from Karl and/or his writings and tapes, as detailed previously.

3. David Armstrong (PhD, Kansas University, 1971) is an ecologist at the University of Colorado, specializing in mammals.

4. See note 1.

5. A. Wetmore, "A Checklist of the Fossil and Prehistoric Birds of North America and the West Indies," *Smithsonian Miscellaneous Collections* 131, no. 5 (1956): 1–105.

6. A. V. Sochava's 1969–1972 publications are considered pioneering classics in the field of eggshell research, especially in Central Asia. For example, see "Dinosaur Eggs from the Upper Cretaceous of the Gobi Desert," *Paleontological Journal* 4 (1969): 517–527 [English edition]; "Two Types of Egg Shells in Cenomanian Dinosaurs," *Paleontological Journal* 3 (1971): 353–361 [English edition]; and "The Skeleton of an Embryo in Dinosaur Egg," *Paleontological Journal* 4 (1972): 527–531 [English edition].

7. Heinrich Erben was also a pioneer in fossil eggshell research. On this topic his writings include "Ultrastrukturen und Mineralisation rezenter und fossiler

Eischalen bei Vogeln und Reptilien" (Ultrastructure and mineralization of recent and fossil bird and reptile eggshell), *Biomineralisation Forschsber* 1 (1970): 1–65.

8. "Which came first, fossil or egg?" *Rockwell News* II, no. 3 (February 13, 1977).

9. R. D. Scheetz and B. B. Britt, "Paleontological Discoveries of James A. 'Dinosaur Jim' Jensen in Central Utah," in *Central Utah: Diverse Geology in a Dynamic Landscape*, Utah Geological Association Publication 36 (2007): 455.

10. Technically, Karl's first paper, published in 1975, was on ammonites [K. F. Hirsch, "Die Ammoniten des Pierre Meeres in den westlicjen USA," *Der Aufschluss*, 26 (1975): 102–113]. His first paper on eggshells was an obscure joint contribution at a workshop [K. F. Hirsch and J. Bowels, "Early Eocene Crane-Like Eggs," *Proceedings of the 1978 Crane Workshop* (Rockport, TX, December 6–8), 212–216]. His first sole-authored paper in a major journal was "The Oldest Vertebrate Egg?" *Journal of Paleontology* 53, no. 5 (1979): 1068–1084.

11. See note 8.

12. This heartfelt exposition on his feelings was written separately on the anniversary of Hildegard's death. It is not part of his longer *Walk in Your Moccasins* manuscript. He also wrote several short expositions under the title "LIFE—Is It Worth Living?" Despite the gloomy titles, these were notes that friends like David had asked him to write in preparation for presentations to students. The objective of such talks was to show the value of overcoming adversity and not becoming self-pitying.

CHAPTER TEN

Eggs, Explanations, and the Shell Game

Origins? Vexing problems in science and philosophy often revolve around origins, and misconceptions about origins. What or which came first, the dinosaur or its egg? We want to understand the origin of the universe, of life, of dinosaurs, primates, and human consciousness. The origins questions for all sciences—astronomy, geology, biology, evolution, psychology, so many others—are complex, as yet unanswered, and perhaps mostly intractable. But we are curious creatures and such questions must at least be asked. They are the same questions asked by the child and by the mature philosophical scientist: Where did I come from? Where am I going? Where indeed are we all going? And while we're asking these questions, we almost unconsciously keep putting one foot in front of the other, as if we had an inkling of where our paths lead.

Did Karl ask these origin and destiny questions as a motherless child? He surely thought them at times when his young mind was, perhaps inevitably, bewildered. And surely he and his soldier comrades asked the big Where are we going? questions in the trench battle lines. There is, they say, nothing like the fear of imminent death to focus the mind, and there are few atheists in foxholes. As a mature man, free from the immediate horrors of war and the upheaval of emigration from the Old World to the New World, Karl had finally grounded himself in a well-defined, mechanical, and technical job, with leisure, much expanded after retirement, to engage in boots-on-the-ground geological excursions. Whether he knew it or not, his search for fossils was, as it was for all people with such paleontological interests, an inquiry into origins, into one or many of the myriad chapters of Earth's history. Karl would do as Hildegard had made him promise to do: soldier on with the next quest.

Ever since Darwin, as biologists like to say, it has been conventional wisdom that any evolutionary inquiry into the family tree of life opens to a lost world of trilobites, ammonites, dinosaurs, and forgotten human ancestors. All are mostly long gone from today's living world, but they are still enshrined in the enduring, mineralized, and near-eternal crust. The planet's shell, which, like our spherical skulls, holds Earth's memories, or like the eggshell, holds the secret of life, the mysterious secret of the "quickening" embryo. What is more, our ancestors are often strangely familiar in appearance, relatives on the tree of life, with heads, limbs, eyes, and symmetry. To pick away at these ancestral remains, mere bones, mere shells of their former selves, is to excavate the physical manifestations of the fossil record. But as soon as we think about their meaning, we enter that very human realm of psychological curiosity about our origins, time, and destiny, and we wonder.

Has any thoughtful scientist or individual not ruminated on the so-called cosmic egg that gave, we are told, birth to the entire universe? And what of the first biological egg from which the first sea creatures hatched and grew to sexual maturity so that they too could produce the next generation of eggs? Which came first: the egg from which the sea species matured, or the individual that created the aptly named "creature"? It's an intractable origins question. Is it even the right question? One book, entitled *The Incredible Egg: A Billion Year Journey*, boggles us with statistics on the trillions of eggs produced year in, year out by countless millions of fecund creatures, from clams and snails to fish, reptiles, and birds.[1]

Karl's interest in fossil eggs is highly symbolic, speaking to the perennial quest of origins. His first obscure paper on eggs, published in 1978, his fifty-seventh year, dealt with crane eggs. The crane, going by the Latin name *grua* or *grus*, has three long toes resembling a trident. Its foot looks like the three prongs used to show family-tree relationships branching down from generation to generation. So much does this three-pronged symbol resemble the bird's foot that since medieval Norman times it has been known as *pied de grue*, from which we get the English word "pedigree"—the very essence of the evolutionary trail of footprints stepping back to our origins in deep time.

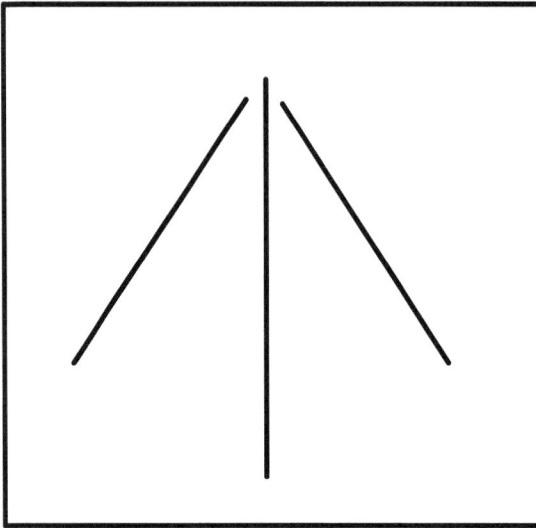

The trident-shaped motif often used to connect generations in family trees.

So here was Karl, on the trail of crane family members that had laid eggs fifty-five million years ago in what would one day be Colorado and Wyoming, in what would one day be the High Plains and Rocky Mountains. Here Karl could watch the iconic Sand Hill cranes on their continental migrations, their free spirits winging over the prairie vastness where he too could roam free.

Karl had a sample of three eggs to work with, including one he himself had found. Knowing they were fifty-five million years old, he dubbed the likely layers of these eggs as "pre cranes." It is, he wrote, "a dream to find a fossil egg with embryonic bones," and his dream had come true with fossil egg number 82, from Wyoming, which contained a half dozen skeletal pieces, including ribs, limb and pelvic bones, and a claw. All could be compared generally, if not in fine detail, with the bones of various species of living crane. Karl's dream embryo was a "baby" prenatal chick, one might say a pre-precrane, an early-stage-of-development embryo, little removed from the origin of its species. Karl was, in his first departure into oology, the science of eggs, deep into the mystery of origins and the arcane world of precrane pedigree.

Like Earth's crust, the eggshell is the strong but brittle exterior of a whole, well-rounded system. Just as the petrified crust encloses broiling, flowing

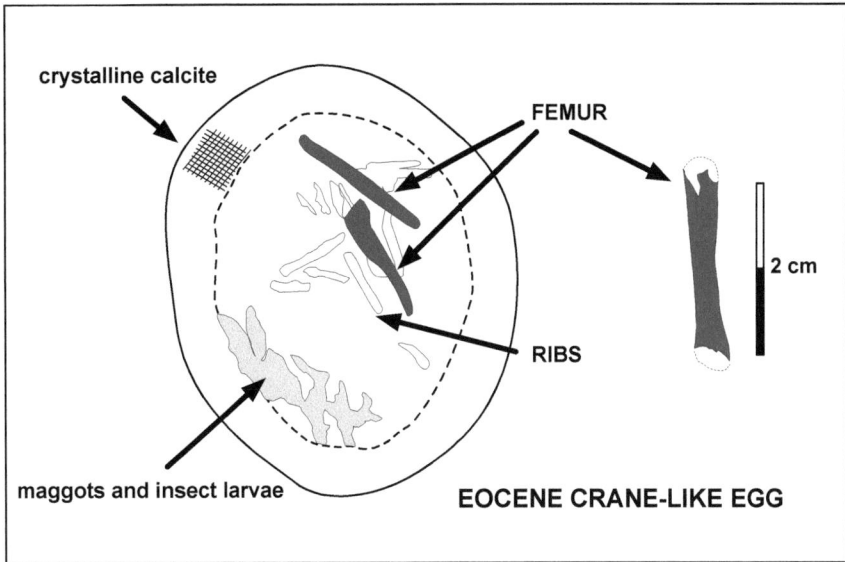

An Eocene crane-like egg with some of the larger rib, pelvic, and leg bones preserved. Note the inner coating of calcite crystals (not original eggshell) and the remnants of maggots and insect larvae. Redrawn from a photograph in a 1978 paper by Karl Hirsch and Jessie Bowles.[2]

magma and lava, organically exhaling its volatile volcanic breath, so the eggshell is the apparently inert or dead, mineralized, sclerotized boundary layer that protects the organic, membranous miracle of life. The shell shields the embryo from desiccating exposure to the rigors of the outer world. The nursery is a sanctuary protected by strong walls that cannot easily be caved in or huffed and puffed down. Young life is fluid, flexible, and dynamic in development, protected by a mineral wall frozen in time. As Karl would later explain, the egg is like a house or home in which life is nurtured (part 4). He had at times known homelessness, something of an occupational hazard for displaced persons. As a youth he could remember interludes of youthful and routine security in a prewar homeland, no longer what it used to be. Lately his mission had been to establish an American home away from home. The egg would become for him a powerful home metaphor, a symbol of stability.

Not every egg is a dream egg. In fact, very few give up their embryonic secrets. The shell that protects them cannot prevent or protect from all earthly hazards. Egg death may come in many guises, the stomp of a clumsy dinosaur, inundation by floods, the desertion of the nest if a parent perishes or is killed by the same predators that would readily eat the eggs. The dream egg is in fact something of a sad story—a healthy egg with an untimely end, grown to within a day or two of hatching before having its life cut short, before being entombed and fossilized for near eternity. Preserved dutifully in its solidly protective shell, its potential transforms from vibrant life to a frozen museum specimen that only a paleontologist could love.

In his second paper on eggs, entitled "The Oldest Vertebrate Egg?," Karl wrote, "The purpose of the eggshell is to protect the embryo and to mediate the interaction between the developing embryo and the outside world."[3] This is the age-old protection metaphor. Parents protect their offspring; presidents, kings, and queens protect their subjects, at least the good ones do. Darwinian evolution is something of a protection racket in which winners, often represented by fiercely protective parents, are said to survive successfully, to live another day, or to, in the case of the egg, help produce the next generation, the next branch on the evolutionary tree. Put your faith and eggs in the basket of the future.

This view gives the eggshell a "mind," a destiny, a species of "purpose" or free will. Sticklers would say the eggshell has only a "function" evolved by natural selection. Intriguing as questions of purpose might be, this was not Karl's point. He spoke to a biological truth that eggshell and embryo are an integrated whole. While the shell may appear inert, mineralized, dead and the embryo fluid, squishy, and alive, they together make a whole entity, symbolic of the paradox of existence that, inside and out, life and death are inseparable polarities that define each other. In short, the shell is not merely mineral. Rather it is, as we shall see, a biocrystalline product, grown by a membrane and shot through with organic material—the interface between the worlds of mineral and tissue.

The eggshell serves a stolid life-sustaining "purpose"—that tricky word again. It is full of little perforations that allow the walls to breath and exchange the gases of respiration: inhalation and exhalation, the eternal and universal rhythm. This dynamism operates beyond the egg, for the intake of oxygen and exhalation of carbon dioxide depend on the partial pressures of these gases "outside" the egg, including in the nest environment. Even the earth's shelly crust is modulated by a similar dynamic cycle of intake (subduction) and outburst (eruption).

Such an analogy is by no means a "stretch." In delving into the story of "The Oldest Vertebrate Egg?" Karl was very deliberately asking about deep-time origins of the eggshell—a science that was, in the 1970s—please excuse the pun—if not at an embryonic stage, at least still in its infancy. Karl's exposition entitled "The Oldest Vertebrate Egg?" has all the marks of a naively wise beginner.[4] This may seem a contradiction since, ostensibly, beginners are inexperienced and not mature, seasoned thinkers. But this is a mere stereotype: the neophyte may have fresh eyes and be better able to see the big picture forest rather than the specialized trees. The new-to-the-field beginner can do no better that review what academics like to call "previous work." This is exactly what Karl did in his second paper. He reviewed the whole field, establishing a platform of useful knowledge from which to launch himself.

His exposition informs us that the egg and eggshell evolved not just as all life has, from once microscopic single-celled creatures into often giant vertebrates like dinosaurs, but in more subtle ways. As paleontologists have learned, there was something of a parallel evolution between the vertebrates and their eggs. This is particularly striking in the evolutionarily significant transition made by vertebrates as they moved from sea to land during the Paleozoic Era of "old life." This sea change was in fact a land change, made as fish evolved into salamander-like amphibians and began colonizing the rivers, lakes, and humid coastal wetland areas where early plants and invertebrates (insects, millipedes, and other arthropods) were taking root and gaining footholds. As their protective skins needed to adapt, so too did their eggs. Early amphibians had, as they still do today, soft porous skins and gelatinous eggs—think of frog spawn, laid in water. However, as various amphibians evolved into various tribes of reptiles that progressively colonized drier interior regions of the continents, so moist porous skin and soft membrane-enclosed eggs were no longer adaptive. Reptiles evolved drier, scaly skin, and their eggs first became leathery and parchmentlike, what Karl would call "soft shelled," then later mineralized by bonelike calcium carbonate, thus becoming "hard-shelled." The latter, more robust types were, of course, more resistant to decay and casual damage and had greater potential to survive as fossils for study by Karl and his colleagues.

When did these first hard-shelled eggs enter the fossil record? As Karl reminds us, it was some two hundred fifty to three hundred million years ago, in the Permian period, before the famous "Age of Dinosaurs" when eggs are found by the dozen or in some places by the hundreds or thousands.

Notes

1. D. Stivens, *The Incredible Egg: A Billion Year Journey*, illus. Bob Hines (New York: Weybright and Talley, 1974), 373.

2. K. F. Hirsch and J. Bowels, "Early Eocene Crane-Like Eggs," *Proceedings of the 1978 Crane Workshop* (Rockport, TX, December 6–8), 212–216.

3. In K. F. Hirsch, "The Oldest Vertebrate Egg?" *Journal of Paleontology* 53, no. 5 (1979), see page 1069.

4. Hirsch, "The Oldest Vertebrate Egg?"

CHAPTER ELEVEN

Light on the Subject

The light which we have gained was given us, . . . by it to discover
onward things more remote from our knowledge.
John Milton, Areopagitica[1]

If you don't have the almost-hatched embryo that makes for a dream egg, all
you have is an empty shell. But there is more to eggshell than meets the eye.
We've already mentioned the little gas-exchange perforations, also known as
pores, which incidentally are small enough to keep out potentially harmful
bacteria. It turns out that like the shells of mollusks and other invertebrates,
eggshell has a complex and varied microstructure, not visible to the naked
eye. One of the standard ways to study this fine structure is to grind down a
little piece of shell until it is only a fraction of a millimeter thick, say a tenth
or a twentieth of a millimeter (thirty to fifty microns, or thousandths of a
millimeter), making what is aptly called a "thin section." These are standard
issue in many branches of science and are so thin they are translucent. Put
one under a microscope and you can quite literally shed light on your subject
and become intimate with the biocrystalline microstructure—make that
simply "microstructure"—of your chosen specimen.

Slicing a piece of biological tissue, or grinding down an eggshell, a fossil
clamshell, or a piece of rock requires a certain technical skill. It is not proverbial rocket science; many graduate students have learned to make thin
sections. But it does take a little patience and skill in what paleontologists
call "prep," that is, preparation of your specimen, to yield the best scientific
results. After Karl had finally settled in Colorado he had, as we have seen,

taken a reliable and secure job as a machinist at Rocky Flats, a secret installation in the shadow of the scenic Rocky Mountain Front Range, notorious for its production of nuclear weapons. Karl's friends questioned, even mildly chastised, him for being involved in the weapons-of-war business, especially after his horrific experiences on the battlefront. He would smile, perhaps chuckle, and decline to address the question directly or enter into the socio-political complexities and ethical ramifications that such a discussion might entail on or around a liberal campus. If nothing else, he had washed ashore in the perfect place to hone the technical machinist skills he had worked hard to add to his résumé. As an apprentice he surely had not anticipated that these skills would serve him well in the scientific eggshell business. But they would also save him the consternation of having to rely on technicians with busy schedules and being faced with the price tags that hung on specialized technical work.

The 1960s aroused questions of the ethics involved in the manufacture of nuclear weapons, as was made all too clear by such social conscience songs as Bob Dylan's "Masters of War" (1963). Robert Oppenheimer, who had been head of the Los Alamos National Laboratory and had become infamously famous for his role in creating the atomic bomb, would famously quote the *Bhagavad Gita*: "I am become Death, the destroyer of worlds." His brother Frank was, at the time Karl moved to Colorado, teaching physics at the University of Colorado. Both brothers, steeped in the discipline of physics, would have been acutely aware of the prevailing wisdom that the universe had begun with a "big bang" and the unleashing of monumental nuclear forces. The fear was rife that human technology would release another big bang that would be a destructive ending rather than a creative beginning.

Links between Karl and the Oppenheimer brothers may seem tenuous, at least beyond their different connections to the Rocky Flats nuclear facility, but the destinies of the father of the atomic bomb and the survivors and casualties of World War II violence were inextricably linked across the Northern Hemisphere, as were issues of geopolitics and conscience. These men would share the experience of being scrutinized by the FBI in order to determine their political and social justice leanings. While a certain conventional wisdom says that the use of the atomic bombs on Hiroshima and Nagasaki brought the war to a quick end, reducing the potential loss of American life, Robert Oppenheimer was regarded by anticommunists as a liberal enemy of conservatives, especially those who still thought like Joe McCarthy. Despite his role as a celebrity who had helped "end the war" and despite his important position with the postwar US Atomic Energy Commission (AEC), Oppenheimer would run afoul of McCarthyism, and in 1954 he would have his security clearance revoked, not without controversy. He had, like many intellectuals of his generation, including his brother, been interested in Communism in the 1930s, when the alternative, Fascism, had hardly seemed attractive. As one Wikipedia entry states, "Oppenheimer was seen by many in the scientific community as a martyr to McCarthyism, a modern Galileo or Socrates, an intellectual and progressive unjustly attacked by warmongering enemies, symbolic of the shift of scientific creativity from academia into the military."[2]

McCarthyism would also torpedo the career of Frank Oppenheimer, who like his older brother Robert had studied briefly at intellectually radical Cambridge University in England, with some of the best minds in physics. Both brothers had been called to testify before the infamous congressional House Un-American Activities Committee. Frank lost his University of Minnesota job as a physics professor and was blacklisted from getting a teaching position in the US. It would be twelve years before Frank was offered a job at the University of Colorado, after he and his brother were rehabilitated to good standing under the Kennedy administration.

As a tenuous reminder of the common experience of Karl and the Oppenheimer brothers, besides at times walking on the same University of Colorado campus, and at times but in different ways contributing to the making of nuclear weapons, we find among Karl's possessions copies of German philosophy books that had belonged to, and been signed by, Robert Oppenheimer.

Karl's nuclear facility job ensured him a decent living involving precision, cleanliness, and personal security, albeit with ominous military implications—a vivid contrast to the insecure messiness of trenches, displacement, and wandering refugee status. So, he was well-placed to become adept at making thin sections, many of which would facilitate his scientific work and feature in the published record. It was "do it yourself" paleontology, supported by "do it yourself" technical assistance, doing what larger geology departments would often hire technicians to do. Starting as he would continue, he would be meticulous in his preparation and study of the *Oldest Vertebrate*

Egg, a priceless one-of-a-kind specimen cherished by Harvard University's Museum of Comparative Zoology.[3] He would succeed in looking into the microstructure of a fossil egg that had guarded its secrets for more than a quarter billion years. The results would be well-cataloged photographs, measurements, thin sections of a specimen that by its very label, "oldest," was paleontologically iconic.

A detailed exposition on eggshell structure would read like a boring instruction manual. As Karl wrote, the "terminology is confusing and in need of clarification."[4] On the other hand, the microscopic world holds much fascination, and there is always more to explore. It is through such study that we know that the soft membrane that encloses the amnion—the white of the eggs—which in turn encloses the nutritious yoke, is what can be called a seeding membrane, on which the shell's mineral crystals grow. First the crystals grow in the shape of a fan, rather like an inverted pyramid, an outward-expanding solid cone, comprising the "cone layer"; then they grow as cylindrical columns, like a neatly packed box of chalk sticks. Like a view of an open chalk box, the ends of the columns can be seen on the eggshell's outer surface, where they protrude as bumps, nodes, or an array of tubercles a few millimeters in diameter and visible to the naked eye. Such bumpy external eggshell ornament is common in dinosaur eggs, as we shall see in chapter 15, but not characteristic of the shells of modern reptiles like lizards and turtles.

A fragment of fossil or modern eggshell can be viewed from either the concave inside or the convex outside. The growth of little cones from the organic, fibrous seeding membrane layer results in the cone layer, which reminded some scientist (a guy, no doubt) of breasts, or mammary glands, leading to their being labeled as the mammillary layer. Terms such as *mammilliform* (breast-shaped) have been widely used in zoology, botany, and even the meteorology of cloud formations. It is perhaps an appropriate term to use in the mineralogical study of crystalline shell formation, for the breast-shaped cone grows from a nipplelike point, which is in a true sense the source of growth and shell creation. It reminds us that, contrary to the simplified notion of an inert mineral shell enclosing a squishy embryo, the shell is in fact a complex organic product of growth, like the bones of a vertebrate. Not only does it develop from a soft inner seeding membrane, it is also enclosed by an outer skin or cuticle.

It is no coincidence that the shell of clams, snails, and other mollusks is quite similarly structured. From a technical viewpoint, then, shells are not really mineral structures enclosing squishy flesh. Quite the reverse, they are created organically, almost miraculously by soft membranes. There is even debate as to how chickens can create calcareous (limey) eggshell even when they ingest no such limey mineral material in their food.

Astronomers tell us the universe and everything in it is expanding. If this conventional scientific wisdom is true, the cosmic egg has expanded considerably since its conception and hatching to release its vast energy and awesome growth potential in the big bang event. Since life first established itself in the form of microscopic cells, most organisms have grown larger, in two ways. As each microscopic egg is fertilized it begins to grow larger, to expand, mirroring a universal law of growth, until it reaches an optimum, full size, and this even before it is laid. In the longer term, beyond the individual's lifetime, the eggs of many, though not all, species have become larger. Thus, the largest, football-sized dinosaur and bird eggs are much larger than those of most of their ancestors. But life is diverse, and while some microbes stayed small other species grew huge, so some egg layers remained small, producing small eggs, while others grew large and laid large eggs. Different eggs grew at different rates and achieved different sizes, reflecting the species that laid them. But as is so often the case in nature, there are exceptions that prove the rule: small species like the chicken-sized kiwi lay disproportionally large eggs, and large species like the otter-sized platypus lay tiny eggs. Likewise, some species lay many eggs, others only a few, in what biologists refer to as different "reproductive strategies." So, "marine biologists once counted the eggs in a 17-pound turbot [fish] and found nine million," a fecundity comparable to that of the common cod fish.[5] Sharks, by comparison, produce only one or two dozen, which they incubate internally sometimes for a couple of years.

The expanding universe is mind-boggling, perhaps incomprehensible, to anyone who thinks too much about it. Equally mind-boggling is the fact that all matter in the universe is not expanding at the same rate. So, any egg or organism that is growing and expanding is doing so at a different rate and to some degree independently from the universe, like a bubble within a bubble, or a cell within a body, growing faster than its container, at least for a while. This is not just a metaphor: it applies to the egg, which, while structurally like a skull, with bony exterior and soft, squishy embryonic interior, expands only slowly but has within it the rapidly growing offspring that is destined to break out of its restrictive skull-shaped crucible. So, what happens when the new generation outgrows the boundaries created by the parents?

The metaphor for an expanding cosmic egg is most appropriate for the study of one of Karl's reptile subjects, the small, four-inch-long zebra-tailed lizard from the southwestern Sonoran and Mojave Desert scrublands of Arizona, California, and northwestern Mexico. This lizard's Latin name, *Callisaurus draconoides*, reminds us that it is seen as a type of "'dragon." One of this species' most distinguishing features is the ability of its eggs to "absorb water and swell considerably during incubation, increasing in mass by as much as 100% during the first half of incubation."[6] Incubation takes about thirty-one days, so the first half of this period is about two weeks, a far cry from the fourteen-month-long incubation period of New Zealand's little, rather sluggish tuatara, a famous living fossil described in chapter 12. Incubation in the deserts' fast lane is helped by an average temperature of 35°C (95°F), well above the comparatively glacial temperature of ~ 11°C (52°F) in which the Tuatara egg develops.

Common sense tells us that if a lizard laid eggs with rigid shells, such a doubling of mass would be difficult if not impossible. Some eggs clearly have their own dynamic ability to expand. They do this by producing at the time of laying (known as oviposition) a very flexible, soft, fibrous membrane. Although crystals grow on the outer surface of the membrane, potentially allowing a little protection, there is a "general lack of organization of the mineral layer" and therefore little structural stability.[7] The growth of the embryo within the expanding membrane may even steal (absorb) calcium from the crystalline layer, preventing it from becoming too rigid and confining. Give and take. In the zebra-tailed lizard the eggshell does not fit tightly around the egg contents, and in fact is organized into a series of wrinkles. Thus, it is like a wrinkled or shriveled prune with a surface that can level out as it swells. Once the wrinkles have been smoothed out, it can expand further as the fibers stretch. Symbolically, indeed literally, as the offspring of the new generation grows, the eggshell expands to accommodate it until

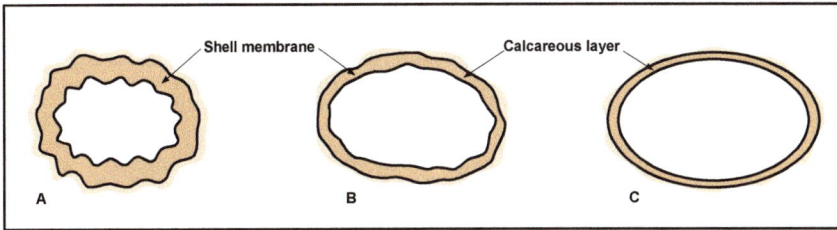

The expanding egg shows nature's foresight in planning for growth of the baby lizard during incubation. Modified from a 1982 paper by Karl and his colleagues.[9]

the "breaking point" is reached. Everything organic grows and expands at some subtle and well-calibrated rate. The egg is capable of growing and expanding, as is the hatchling that emerges and grows to adult size, while the universe is itself expanding. Is it true to say the expanding egg is found in the expanding mother's oviduct, while the whole ecosystem and biosphere in which they are contained expand simultaneously? These are questions to intrigue philosopher scientists and biologists alike. For a more awe-infused view of the remarkable, expansive properties of the growing reptile, we have the British naturalist Gerald Durrell's wish to "fathom the mystery of how a baby tortoise, emerging from its egg as crushed and wrinkled as a walnut, would, within an hour, have swelled to twice its size and have smoothed out most of its wrinkles."[8] Without saying so explicitly, Durrell had noticed the hatchling turtle going very rapidly through the same dynamic and expansive development from wrinkled and shriveled to enlarged, swelled-out phases, as has been noted in certain eggshell development. Such resonance of shell and hatchling development speaks to the intricacy of biological organization.

To the human way of thinking, flexibility is a valuable adaptation (as is flexible thinking), but so is strong protection from environmental dangers. The benefits of both the strategy of strength and that of flexibility are very nicely summarized by Karl in his 1983 study of turtle and tortoise eggshells.[10] Not all turtles are created similar or equal. There are sea turtles, which lay their eggs on land—that is to say, buried in sandy beaches; freshwater turtles, such as the famously bad-tempered snapping turtle; and tortoises, such as the equally famous but patient and apparently good-tempered Galápagos giant tortoise. We should perhaps not offer any big prizes for guessing which of

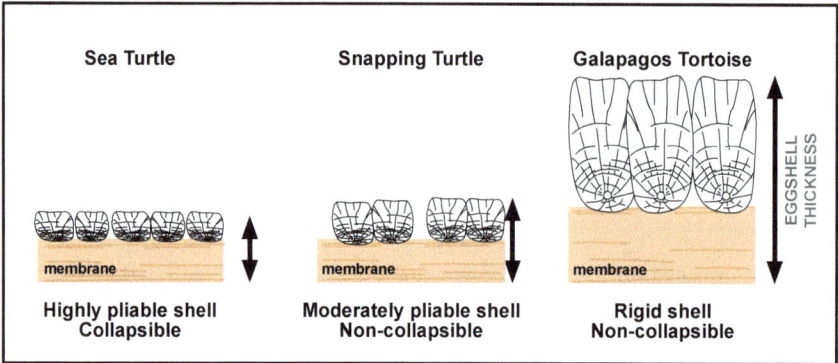

Sea Turtle	Snapping Turtle	Galapagos Tortoise

Nature is ordered, as shown in turtle eggs. Sea turtles have thin, pliable, and collapsible eggshells, dominated by soft membranes. Freshwater turtles like the snapping turtle have slightly thicker, less pliable, and noncollapsible shells with proportionally thicker shells and thinner membranes. Terrestrial tortoises have the thickest, most rigid eggshell and thinner membranes. Diagrams modified from a 1982 paper by Karl Hirsch.

these three types has the thinnest, most pliable and collapsible eggshell; the intermediate, moderately pliable and noncollapsible eggshell; or the rigid, noncollapsible eggshell.

You surely guessed correctly. A simple diagram shows us the difference. The collapsible sea turtle eggshell has a membrane one and a half times the thickness of the flimsy crystalline layer. By contrast, the Galápagos giant tortoise has a rigid crystalline layer twice as thick as the membrane. This is a fine example of a biological polarity, a well-organized give-and-take regime, representing a balance between the squishy organic membrane and the bio-crystalline shell, which is manifest across the turtle-tortoise family tree. Put another way, the wise turtle clan have learned the advantages of both flexibility and strength and have conducted evolutionary experiments accordingly.

Notes

1. This John Milton quote is taken from his book *Areopagitica*, published in 1644. Its message is that enlightenment deepens our knowledge.

2. See "Oppenheimer security hearing" on the Wikipedia website (https://en.wikipedia.org/wiki/Oppenheimer_security_hearing). Originally quoted in

R. G. Hewlett and J. M. Holl, *Atoms for Peace and War, 1953–1961: Eisenhower and the Atomic Energy Commission* (Oakland: University of California Press, 1989), 110.

3. In K. F. Hirsch, "The Oldest Vertebrate Egg?" *Journal of Paleontology* 53, no. 5 (1979), see page 1069.

4. Hirsch, "The Oldest Vertebrate Egg?"

5. D. Stivens, *The Incredible Egg: A Billion Year Journey* (New York: Weybright and Talley, 1974), 81.

6. In M. J. Packard, L. K. Burns, and K. F. Hirsch, "Structure of Shells of Eggs of *Callisaurus draconoides*," *Zoological Journal of the Linnean Society*, 75 (1982): 297–316, see p. 298.

7. Packard, Burns, and Hirsch, "Structure of Eggs," 313.

8. G. Durrell, *Fauna and Family: A Nonpareil Book* (Boston: Godine, 1978), 37.

9. M. J. Packard, L. K. Burns, and K. F. Hirsch, "Structure of Shells of Eggs of *Callisaurus draconoides*," *Zoological Journal of the Linnean Society* 75 (1982): 297–316.

10. K. F. Hirsch, "Contemporary Fossil Chelonian Eggshells," *Copeia* 2 (1983): 382–397.

PART 4

Man of Letters

CHAPTER TWELVE

Dragons, Dinosaurs, Maternity, and the Dinosaurian Big Time

The tuatara, a New Zealand lizard look-alike, but not a true lizard, two feet long when fully grown, is much celebrated by paleontologists, despite being a rather sluggish creature. The reason for this fame is pedigree. The tuatara, Latin name *Sphenodon punctatus*, has been dubbed an "ancestral lizard" with relatives that lived in the Triassic period at the beginning of the Age of Dinosaurs more than two hundred million years ago. Biologically the tuatara, a Maori name referring to its spiny back, belongs to the reptilian order Rhynchocephalia, mostly known to paleontologists due to all its members other than the tuatara being extinct! It nevertheless probably shared common ancestors with true lizards dating back to the dawn of the dinosaurian era.

So, investigating the tuatara egg, as Karl did in the early 1980s, may shed light on the evolutionary origins of lizard eggshell. Biologists and paleontologists are lucky the tuatara is still in the land of the living. Having evolved very little, or very slowly, in the last quarter billion years since the Triassic, it is what paleontologists call a "living fossil." This apparently contradictory if not confusing and contentious term describes species that appear "primitive," having evolved very little compared to other related species. The differences highlight polarities between conservative and progressive evolutionarily strategies. Despite being rather sluggish, the tuatara has a conservative, "go slow" metabolism that has helped it survive where others could not. Like turtles, which have tried out the diverse and perhaps equally beneficial strategies of flexibility and strength, lizards and other reptiles evidently experimented with both hot and cool blood as well as slower and faster evolutionary rates.

The tuatara's luck in the survival game has much to do with the species' isolation in both geological and historical time. This "little dragon," as it has sometimes been called, has many remarkable adaptations. Not least of these is a low body temperature ($52°F = 11°C$) and a remarkably drawn-out incubation period. The female lizard lays eight to fourteen soft-shelled, parchment eggs in the spring, burying them in the soil, where the embryos develop slowly through the summer. In the autumn and winter, their development slows down as they (and adult lizards as well) go into a type of hibernation. The eggs don't hatch until the following summer, after thirteen to fourteen months of incubation without parental care. The hatchlings have a horny egg tooth on the end of their snouts to help break the tough, leathery eggshell; this snout or beak hatching device is common to many species, including birds. The first rays of light to illuminate the young lizard's head are symbolic of another extraordinary feature of tuatara anatomy: the light-sensitive pineal gland or "third eye" that is also present in fish, most amphibians, and some extinct reptiles, an anatomical feature with deep time evolutionary roots. In the hatchling this gland is covered by semitransparent skin, rather like a thin section, but in the adult it is covered by opaque scales. The newborns, it seems, manifest an ancient enlightenment potential that has been lost in the adults!

Since the age of dinosaurs, New Zealand has been geologically isolated terrain. Because this isolation has protected it from many of the most competitive and rapacious predators that evolved on mainland continents, many island species, notably large flightless birds, evolved in comparatively pressure-free safety, as did the tuatara. Terrestrial mammals never gained a foothold in New Zealand, even though they became the dominant class in the rest of the world. Sadly for species like the tuatara, however, living in comparative safely, all this would change in very recent historical, not geological, time, when a dangerous new species, *Homo sapiens*, appeared on the scene. The first arrivals were the Maoris, a human Polynesian culture that was very familiar with the Pacific Islands and their often-scarce animal resources that provided only a limited supply of protein. It is well known that the Maoris encountered a dozen species of large birds collectively known as Moas, the largest of them resembling gigantic ostriches with giant eggs to match. Like ostriches, these had become flightless, due to lack of predation pressures, and they also lived in mostly wooded habitats with no need for the savannah speed evolved by ostriches. The Maoris essentially killed them all before the first Europeans arrived, perhaps paying little attention to the inconspicuous tuatara. Campsite archaeology showed that they did apparently kill and presumably eat the little dragon from time to time, but it seems unlikely they gorged themselves on mountains of scaly, reptilian flesh, as they had evidently been able to do with Moa flesh. However, the tuatara would encounter another equally dangerous and difficult-to-control predators: rats. These large rodents, famous for breeding successfully almost everywhere, have been the ecological scourge of countless islands around the world. Today the tuatara survives on only a few rat-free offshore islands and a few other islands where the rat and feral cat populations have been kept under control. In many places the little dragon shares its favored burrows with petrels, smallish members of the albatross tribe, on which it sometimes preys by taking eggs and chicks—when it is not eating the crickets and other insects that make up part of its diet.

Now that we've "shed light" on the famous tuatara's present-day ecology, geological history, and precarious future, what of its eggs? Karl, of course, did not raid a little dragon clutch laid by this endangered species for his research specimens. Instead, he obtained, from the New Zealand Wildlife Service, small inconspicuous shell fragments from eggs of juveniles already hatched. Karl would focus much scientific equipment and power on this little bit of evolutionary parchment. To quote his paper directly: "The structure of the shell fragments was studied by scanning electron microscopy (SEM), polarizing light microscopy (PLM), X-ray diffraction (XRD), X-ray fluorescence (XRF) and X-ray spectroscopy (XRD)."[1] These "X-file" methods involve large, complex machines that can probe the structure and composition of miniscule particles. By the 1970s and 1980s, a proverbial "arsenal" of powerful analytical tools could be brought to bear on most objects of interest.[2] Karl would systematically employ these tools to extract the eggshell's secrets. These powerful tools worked almost magically to see into the normally invisible physical and chemical structure of matter. Together with his second eye, his microscope, it was as if Karl had access to his own third, fourth, fifth, sixth, and seventh eyes.

Curiosity demands to know: Employing such sophisticated equipment, what results did Karl and his coworkers obtain from their analysis of these little eggshell fragments from halfway around the world? The visually most spectacular results gracing the beautifully illustrated paper, published in conjunction with coauthor Mary Packard and a colleague from New Zealand, show the intricate relationships between the inner organic membrane and the outer calcareous (calcite) shell. The membrane—far from being a soft, smooth, and slimy sheet, as one might imagine from its name—appeared, when magnified two hundred to one thousand times, like a multilayered mat of hairlike fibers, for all the world resembling the mat of hairs caught on the grill over one's shower drain. The membrane is described by these authors as a "tightly woven mat of relatively small fibers,"[3] differentiated into several "layers" with fibers of different size and weave.

We also learn from the small print that this woven membrane plays a role in the egg's ability to swell during incubation, as discussed by students of the tuatara as early as 1899—albeit biologists without the third to seventh eyes and without the X-ray vision tools available to Karl and his co-seers in the 1980s. The swelling or contracting of the egg depends on the humidity in the nest and relates to a factor known as water balance. It is another dynamic of the egg as a living, organic entity. It is all part of the absorbing detail surrounding the reproductive and evolutionary biology of reptiles, their eggs, and eggs in general. No wonder Karl became absorbed himself.

Scientific caution usually requires that conclusions be reached tentatively. The woven inner membrane could expand a little but, due to the more rigid outer calcareous shell, perhaps not as much as in some lizards. Who knew eggs have a potential swell index? Overall, the verdict is that the tuatara egg falls in the range of flexible, soft-shelled egg types—somewhere between the softer shells of turtles and the more rigid eggs of geckos. Whether the tuatara egg bears the signature of "evolutionary antiquity" attributed to the species as a whole, Karl and his friends declined to venture.[4] They had at least probed some of the species' hidden secrets, penetrating beneath and into the very skin of its reproductive membranes.

Writers on the subject of evolution like to quote the final paragraph of Darwin's *The Origin of Species*, where he spoke of "endless forms most beautiful and most wonderful [that] have been, and are being, evolved." This almost poetic statement shifts us from the technical world of anatomy and physiology into one of wonder, where we are reminded that every species is different and a marvel of creation: that is the creative power of evolution. But the message of evolution is also that each "separate" species is evolutionarily connected to more or less close relatives in what was once simplistically called the hierarchical "Great Chain of Being," or if you prefer, as do most modern biologists and paleontologists, as branches on the more nuanced and integrated evolutionary concept of a "family tree."[5]

The next subject of Karl's eggshell research would be in the world of crocodiles, familiar monsters that are, in the public eye, both primitive and dangerous. As reptiles they are related to groups Karl had already studied: turtles, lizards, and other ancestors. Proving their own ancient pedigree, crocodiles are members of a group called archosaurs ("ancient reptiles"), most closely related to their archosaurian cousins, the extinct pterosaurs and the dinosaurs—the latter still represented today, as most schoolkids now know, by birds. If we think of crocodiles as today's archosaurs of the water, and pterosaurs as yesterday's archosaurs of the air, then most dinosaurs were the archosaurs of land habitats. This simple threefold scheme suggests that crocodiles are the only successful survivors. But "not so fast!" says the attentive third-grade biology student. In recent decades we have learned that one group of dinosaurs, the carnivorous "theropods," did not become extinct; some evolved into birds. This just proves our initial thesis that all species are connected. Crocodiles are archosaurian cousins of dinosaurs, some of which are cousins of birds. All of them, not surprisingly, laid eggs that shared various similarities but were nevertheless different from one another, just as they are different from those laid by turtles and lizards.[6]

As implied here, these differences aren't just between today's living reptiles, crocodiles, turtles, and the closely related lizard and snake families; these differences changed through time. Just as today's avian dinosaurs, from birdlike sparrows to seagulls, differ from nonavian dinosaurs of the Jurassic period, like *Allosaurus*, and those from the Cretaceous period, like *Tyrannosaurus rex*, so today's crocodiles differ from those of yesterday. Some early crocodiles were greyhound-like and land-based, like the aptly named *Terrestrisuchus* (terrestrial crocodile; *suchus* means "croc-like"), which weighed a mere thirty pounds. By contrast there was the later, Cretaceous, humongous "terrible croc" *Deinosuchus*, favorite of monster-lovers, with a five-foot-long skull and a body length up to thirty-five feet (about ten meters,) and thus half as big again as the largest of today's saltwater crocodiles.

Karl did not have to go to New Zealand or some tropical river to find his crocodile egg. It had been found in Colorado in the late 1950s, around the time Karl had immigrated to the Wild West. Today, with its freezing winters,

Colorado is hardly croc country. But forty-five to fifty-five million years ago, not too long after the so-called dinosaur extinction sixty-six million years ago, Colorado and even more northerly Wyoming were subtropical. The world had been a greenhouse sixty-six million years ago when the last nonavian dinosaurs, pterosaurs, and various marine families went extinct, thanks, most paleontologists believe, to a huge meteorite impact, for which ever-more-convincing evidence is steadily accumulating.[7] After what may have been a geologically brief cold snap, the surviving crocodiles inherited a still-warm world that cooled down slowly over the next sixty to sixty-five million years, leading to the very recent "Ice Age," which chilled planet Earth intermittently during only the last two million years.

So, forty-five to fifty-five million years ago, during the Eocene epoch (*Eo-cene* means "dawn" of recent times), the subtropical, junglelike flora of Colorado and Wyoming had not given way to today's dry prairie grasslands. Among the many details that Karl and his colleagues divulge about the crocodile eggs from Colorado is that like modern crocodile eggs they have "no resemblance" to turtle or dinosaur eggs and "can easily be distinguished" from those of lizards.[8] On the other hand, they "cannot always be separated" from bird eggshell, at least not with the ordinary light microscope. Both generally have smooth eggshells—in the case of crocs, not unlike those of the Mississippi alligator.

As we shall see, Karl's eggshell collection would eventually be bequeathed to the University of Colorado Museum. It was from the same collection that his croc eggs were taken for study. The best example was one of four from a clutch, and thus evidently still in its nest when found. It is about two inches (five centimeters) long and one and a quarter inches (three centimeters) wide, so about the size of a hen's egg. Like fossil footprints, which are made exactly where an animal trod, fossil nests tell us precisely where a species laid its clutch. Clues in the sediment suggested to Karl and his colleagues that the nest was situated near a river, where the parents had buried the clutch in sand or vegetation, as crocs do today. Eggshell fragments were found at three other sites within a mile of the nest, in some cases with teeth and other remains of crocodile skeletons. Noting that there is a relationship between the size of animals and their eggs, the study concluded that the egg layer was between about three feet and six feet (1.5–3.0 meters) long.

It is important to note that one cannot identify the egg layer "unless identifiable embryonic remains are found within" the eggs or, as more recently
discovered, in the egg layer's body cavity.[9] As noted below, this usually requires researchers to confess that perhaps the egg layer is still unknown:
what we might call necessary frankness or scientific disclosure. Thus, it is
necessary for specialists to come up with appropriate scientific names for
the eggs themselves. This distinguishes the eggs from any known species of
actual or potential egg layer. Tracks can also be named scientifically, without
knowing which species were the trackmakers, and the same goes for fossil
feces (poop)! This is all because it is quite possible that an egg layer or a
trackmaker really is not known from an actual body fossil. In theory, such
egg layers or trackmakers may never be known. Eggs and tracks are first
and foremost only clues to the existence of a species that may or may not be
known from other fossil evidence. As in detective work, the true egg layer or
trackmaker "culprit" or perpetrator can be positively identified only when
certain standards of evidence are met beyond a reasonable doubt—essentially, when we find identifiable embryos still inside eggs.

So, what label did Karl and his colleagues put on the forty-five- to fifty-
five-million-year-old egg from western Colorado? They named it *Krokolithus
wilsoni*, meaning "egg of a croc found by Herman Wilson." Krok needs no
translation, and *olithus* literally translates as "egg rock" (or "lithified egg"),
a term that had been coined, and spelled *Olithes*, by an egg researcher, or
"oologist," in faraway China.[10]

It has been said that a thing does not really exist until it has a name. In
this study Karl and his friends stressed that literature on fossil crocodilian
eggs was sparse and that comparatively little was known about them. The
Colorado study gave a name and scientific standing to a rare fossil, the first
of its type named from Colorado, and the first of a crocodilian to receive such
paleontological recognition. In the grand tradition of naming and classifying
species (which began with Carl Linnaeus, the eighteenth-century so-called
father of taxonomy), *Krokolithus wilsoni* became immortal, like other named
"species" whose names are inscribed in literature, no matter how obscure.
On the subject of obscurity, or more fairly, necessary scientific vocabulary,
the arrangement of species names, or *taxonomy*, literally means ordering
("taxis") of names ("nomy"). For everyday purposes, *nomenclature* has much

the same meaning as *taxonomy*: note the "nomy" root. *Nomenclature* is ubiq-
uitous in science and language, and *taxonomy*, as extended to eggs and tracks,
is sometimes referred to as *parataxonomy*, which simply means "toward taxonomy." Like Linnaeus and his orderly contribution to biology, Karl's little
petrified *Krokolithus* egg will live forever in the annals of paleontology, and
perhaps a future paleontologist will find an egg with that rare, elusive embryo
that will tell us the egg layer's identity and flesh out the egg's pedigree.

When Arthur Lakes first began fossil-hunting in the foothills of the Rocky
Mountains near Denver, Colorado had just become, in 1876, what came to be
known as the Centennial State. Little did this well-educated English clergy-
man, with his Victorian Era passion for Earth history, suspect that his discoveries of Jurassic dinosaur bones would transform the world of paleontology.
He also could never have imagined that the little settlement of Morrison,
situated where Bear Creek flows out of the mountains onto the High Plains,
not far from where Karl would live, would one day give its name to one of
the world's most famous rock formations.[11] The Morrison Formation is to
dinosaur paleontology what Olduvai Gorge is to anthropology or Pompeii
to Roman history.

Following his discovery of giant dinosaur bones attributable to long-
necked, long-tailed sauropods, popularly known as brontosaurs, Lakes would
tirelessly excavate some of the world's most iconic dinosaurs, including *Diplodocus, Stegosaurus*—today the Colorado state fossil—and *Allosaurus*, the
top Jurassic carnivore. At the time, there was little evidence to suggest that
paleontologists thought about dinosaur reproduction, eggs, nests, or hatchlings. It would be a century before such topics became of much interest,
especially in the Jurassic period, or in North America.

The name of Morrison, however, was destined for lasting fame. Arthur
Lakes would camp there, by Bear Creek, using the little hotel in the dead
of winter, while he and his helpers laboriously dug and chiseled bone from
Jurassic sandstones and shales. These they packed off to Othniel Charles
Marsh, the imperious and ungrateful professor of geology at the Yale Pea-
body Museum. It was no small project. After two years of hard work, Lakes
had shipped dozens of crates back East.

His discovery sparked what has been likened to a dinosaur gold rush, a
series of bone wars between those working for Marsh and those working for

his greatest rival, Edward Drinker Cope of Philadelphia (who ironically, as a teetotaling Quaker, was no drinker). By the early twentieth century, the bone-rich Jurassic Morrison Formation had yielded dinosaurs elsewhere in Colorado, Wyoming, and Utah, notably at the site that would become Dinosaur National Monument. By the mid-twentieth century, geologists had analyzed the composition and distribution of the bone-rich strata, defined its upper and lower limits, and named it the Morrison Formation. The little settlement, where Lakes had pitched his tent, now gave its name to a suite of sedimentary rocks that have been identified from Colorado south and west to New Mexico and Arizona, and north to Wyoming, Montana, and the Canadian border. Today we know that the rocks are about 150 million years old, representing the Late Jurassic epoch.

Where egg-laying dinosaurs existed, there is at least a theoretical chance that one may find eggs, nests, or babies. But this does not mean they are either easy to find or easy to identify. In contrast to a large dinosaur leg bone, up to six feet long and a foot in diameter, with its spongy texture heavily mineralized, a fragment of eggshell may be no more conspicuous than a discarded fingernail. When Karl first got into the Jurassic egg hunt, it would be in the Morrison Formation, in his home state of Colorado, and he would be dealing with the first eggshells ever reported from the Jurassic. The material was indeed scarce. One of the three eggshell sites, although dinosaur-rich, had yielded only "a single shell fragment," less than a millimeter in thickness and described as "birdlike." The report was published in 1987, but definitive evidence that the eggs were dinosaurian was not established.[12] Paleontology had waited 150 million years for "Who laid it?" revelations but would have to wait just a little longer.

Publishing in *Science* or *Nature*, the two most prestigious and widely known international science journals, the former American, the latter British, has long been, for those with such ambitions, akin to achieving Olympian status and membership in an elite science club. In rare cases a researcher might just get lucky and stumble on that once-in-a-lifetime, *Science*-worthy find. Accomplishing such a publication coup means the science industry actually confers status and even awards a points ranking, if you care to look such things up, in the *Science Citation Index*. These two journals are read regularly by scientists, science watchers, and science journalists of all persuasions: well informed intellectuals who like to "keep up." Before the days of electronic downloads, a curious dentist in Florida might ask a paleontologist in New York for a copy of an interesting article on dinosaurs. Editors might deliberate for months before selecting noteworthy paleontology articles to shuffle in among competing articles on genetics, medicine, astronomy, and other disciplines. But if you win what aspiring authors regard as a scientific lottery, you are guaranteed a wide audience and at least fifteen proverbial minutes of fame, as your article basks in a warm glow of special and significant interest. Your otherwise taciturn colleagues may even take interest, at least briefly! But don't hold your breath: more than 95 percent of submissions are rejected, and that rate is increasing as the flood of incoming submissions increases.

A Jurassic egg found in a remote spot in the Morrison Formation of Utah would, in 1989, win the *Science* sweepstakes.[13] This time, the discovery was no mere fingernail-sized fragment, but a whole egg, a little distorted by 150 million years of geological faulting and folding pressure, but nevertheless easy to reconstruct as measuring about 11 centimeters long and 5.5 centimeters wide, about the size of the egg of a California condor, one of today's largest living birds. So, who laid it? A strong clue comes from where it was found, in the middle of a well-known dinosaur quarry, named the Cleveland-Lloyd Dinosaur Quarry, famous for a huge and jumbled accumulation of bones of the carnivore *Allosaurus* (incidentally, named the Utah state fossil in 1988). Jim Madsen, one of Karl's coauthors and a former Utah state paleontologist, had devoted much of his scientific energy to studying this iconic carnivore species, *Allosaurus fragilis*, and helping make it famous. Since the first excavations in the 1920s, more than ten thousand bones have been recovered, 70 percent of which belong to *Allosaurus*, including some that represent the smallest individuals in this species. Thus, Jim had abundant material available from which to put together entire skeletons, from which he also made and sold replicas for museum display.

It is unusual to find such large concentrations of carnivore remains when we know that herbivorous prey species usually significantly outnumber predator species, at least in modern ecosystems. For this reason, paleontologists have inferred that this deposit probably represents a predator trap, rather like La Brea Tar Pits, in Los Angeles, only very much older. Instead of La Brea's abundant mired wolves and eagles, the Utah site was probably a theropod dinosaur death trap. In such traps animals became mired in sticky mud when they came to drink or prey on other species drawn to a life-giving, but potentially death-dealing, waterhole. Other theories suggest the quarry is evidence of a drought and the last phases of activity around a dried-up water source.

But why was a single egg found in the middle of this evidence of death? A *Science*-worthy clue comes from the shape of the egg itself. It was, to quote Karl and friends, "broken open, with two halves connected along a hinge like folded area."[14] This was not folding due to geological forces, but rather evidence that the egg was soft, pliable, and "foldable" at the time it became fossilized. It seems likely this could only mean it was still in the oviduct, as yet unlaid, of a gravid mother. Once laid, rigid eggshell of this type would not have remained pliable for more than a few minutes and may even have already hardened in the mother's oviduct before being laid. Two other fascinating clues help confirm this story. First, the eggshell is pathological, meaning that it has an extra, pathological layer laid down on top of the normal layer. This indicates that the egg was retained in the oviduct, where a second unwanted layer of shell was deposited. As Karl and his friend put it, "The retention of eggs and formation of pathologic eggshell . . . is known . . . due to stress, illness or environmental conditions."[15] So here were two lines of evidence for egg retention. In many modern reptiles and birds, the embryo develops for about two days before the egg is laid. A CAT scan of the egg revealed a shadowy object about 3.8 centimeters (one and a half inches) long inside. This seems to be a consistent third line of evidence for egg retention by the mother. Even if the egg had been laid, which seems unlikely, its chances of normal incubation were effectively zero. Not only was the mother apparently in a death trap but also the double-layered pathological eggshell would have caused the embryo to suffocate, because the extra shell layer closes off the pores that allow the egg to breath. Even if by a miracle the pores were sufficiently abundant and aligned, in both layers, to allow respiration, the hatchling would face the challenge of breaking out of an extra-thick shell.

With admirable caution the *Science*-worthy paper, entitled "Upper Jurassic Dinosaur Egg from Utah" does not claim to have proven the identity of the egg layer, although Karl and colleagues tell us it was not a sauropod, which would have had a different shell structure. They do, however, proudly and justifiably claim that "whole dinosaur eggs, pathological eggshell and embryonic remains are extremely rare." If *Science* had allowed it, Karl and his friends might have entitled the paper "Stressed Allosaurus Mom Caught in

A smiling Karl with his eggs in the early 1990s, when he was at the top of the academic game.

Death Trap, Can't Lay Egg, Family Line Terminated." Perhaps this is why we authors are not the editors of *Science* magazine!

In Buddhism, ten thousand is a symbolic "large" number signifying the uncountable, myriad wonders of creation. It is not a number to be taken too literally. After our Utah brethren had counted ten thousand dinosaur bones, quite a large and more-or-less precise number, a single egg came to light. Surely Karl was right that the folded pliable egg, held back in the mother's oviduct, was by any standards "rare."[16] The exception that proves the rule. It had held itself back in the oviduct as well as from early discovery, as so much paleontological evidence often does, until the time was right for observant science to shed new light on a rare treasure. It was almost the case that Karl and friends had stumbled on a whole new line of evidence for identifying egg layers, which would be verified by future studies.[17]

Notes

1. In M. J., Packard, K. F., Hirsch, and V. B. MeyerRochow, "Structure of the Shell from Eggs of the Tuatara, *Sphenodon punctalus*," *Journal of Morphology* 174 (1982), see page 197.

2. Karl could not have done much of the detailed work on eggshell microstructure had it not been for the invention of the electron microscope, which did not come into widespread use until the early 1970s. In this regard, his embarkation into eggshell science was well timed.

3. See Packard, Hirsch, and MeyerRochow, "Structure of the Shell," 201.

4. As explained earlier in this chapter, the somewhat simplified notion of a "living fossil," or evolutionary antiquity became widespread in popular treatments of paleontology, and it generally refers to species that show little sign of evolutionary change over long periods of time. The tuatara is a classic example, as is the coelacanth, a fish long thought to be extinct until discovered in east African waters in 1938.

5. "The Great Chain of Being" or "Scale of Nature" (*Scala Natura*) is an ancient concept dating back to Plato and Aristotle. It recognized the hierarchical organization of life and matter from the highest (God and angels) to humans, higher animals, plants, and minerals. Modern paleontology has recast this concept in terms of progressive evolution and the family tree, still maintaining the notion of a hierarchy of increased complexity (without invisible angels or God). For a thorough treatment, see J. D. Archibald, *Aristotle's Ladder, Darwin's Tree: The Evolution of Visual Metaphors for Biological Order* (New York: Columbia University Press, 2014).

6. It is interesting to note that Karl's early participation with the University of Colorado Museum often involved collecting fossil mammals, which of course did not lay eggs. In finding and studying eggs, he therefore necessarily veered into a research field entirely new to the university. This led him through various reptile and bird groups to ever-popular dinosaurs. Again his timing was good, as dinosaur research enjoyed a marked renaissance in the 1970s.

7. Much celebrated discoveries at the Cretaceous-Paleogene boundary in North Dakota suggest that evidence for an end-Cretaceous impact is irrefutable. See R. A. DePalma et al., "A Seismically Induced Onshore Surge Deposit at the KPg Boundary, North Dakota," *Proceedings of the National Academy of Sciences* 116, no. 17 (2019): 8190–8199.

8. In K. F. Hirsch, "Fossil Crocodilian Eggs from the Eocene of Colorado," *Journal of Paleontology* 59 (1985): 531554, see page 533.

9. Hirsch, "Fossil Crocodilian Eggs," 541.

10. This researcher, Zhao Zi-Kui, is introduced in chapter 15.

11. The Jurassic Morrison Formation, named after a little town nestled in the Colorado Front Range just west of Denver, was practically in Karl's backyard. Karl had collected dinosaur bone from this formation near Moab. Today the famous Morrison

Formation fossil beds are part of the Morrison-Golden Fossil Areas National Natural Landmark, better known as Dinosaur Ridge.

12. K. F. Hirsch, R. Young, and H. J. Armstrong, "Eggshell Fragments from the Jurassic Morrison Formation of Colorado," in *Paleontology and Geology of the Dinosaur Triangle, Guidebook*, ed. W. Averett (Grand Junction: Museum of Western Colorado, 1987), 7984.

13. K. F. Hirsch et al., "Upper Jurassic Dinosaur Egg from Utah," *Science*, 243 (1989): 17111713.

14. Hirsch et al., "Upper Jurassic Dinosaur Egg," 1712.

15. Hirsch et al., "Upper Jurassic Dinosaur Egg," 1712.

16. Hirsch et al., "Upper Jurassic Dinosaur Egg," 1713.

17. Since the 1980s, when it was assumed that one could identify an egg layer only from eggs with identifiable embryos (Hirsch, "Fossil Crocodilian Eggs," 541), eggs have been found in the body cavities of egg layer skeletons as reported by T. Sato et al., "A Pair of Shelled Eggs Inside a Female Dinosaur," *Science* 308 (2005): 375. We are grateful to Darla Zelenitsky, coauthor of this paper, for pointing out to us (the authors) in a written communication that Karl "would have been thrilled" by this new means of identifying egg layer species. We are also indebted to Darla Zelenitsky for noting that, after Karl's death, discoveries of eggs with embryos confirmed his suspicions that the Utah eggs were allosauroid.

CHAPTER THIRTEEN

O is for Ovum, Oeuf, and Ooh!

A little nonsense now and then is relished by the best of men.

Anonymous

The Greek word for *egg* is "öon," from which we get oology, the study of eggs; oologist, the student of eggs and the name of an egg journal; not to mention oogenesis for egg development and oogonia, oopods, oosperm, oospheres, and ootype, all pertaining to reproduction. Oographs and oometers help you measure egg shape, and ooscopes help you look inside. All this "Ooh" vocabulary is even before we get to fossils and what are called "oospecies."

The first oospecies was named *Oolithes bathonicae*, meaning "stone egg from the town of Bath in England."[1] It was named in 1859, which happened to be the year that Darwin's *On The Origin of Species* was published, and also the year that one Father Jean-Jacques Pouech, a French priest and geological enthusiast working in the south of France, found what would later prove to be the first confirmed discovery of dinosaur eggshell. The Bath egg, as Karl later proved, was that of a turtle. The 1859 date was a triple coincidence of timing, perhaps, but not surprising in what was one of many nineteenth-century heyday firsts for natural history, geology, and paleontology.

If you prefer Latin, begin with *ovum*, plural *ova*, and you have ovaries, oviducts, oviparous (egg-laying), oviposter, and ovulation. You might even call egg white "oviplasm" and like your oviplasm over easy. Terms like *oval*, which is even the name of a famous British cricket ground, pop up regularly in everyday speech. Some dictionaries define *ovulite* as a fossil egg, but this is not a term used by Karl and his oologist colleagues. The yolk and white (albumin) of eggs derive their names from their color. Albumin is from the Latin *albus* for "white," and yolk comes from the Middle English *yolke*, meaning "yellow."

As we have just seen, you cannot run around claiming to have an *Allosaurus* or *T. rex* egg without convincing evidence. So, with necessary caution paleontologists—make that paleo-oologists—give specialist, technical names to different types or species of eggshell. For example, one of the three Colorado eggshell sites, the "Young Egg locality," was named after Bob Young, not because the site was geologically "young" but because Bob Young had found it. One of the eggs first named was *Prismatoolithus coloradensis*."[2] This wordy mouthful, which Karl coined in 1994, loosely and somewhat obviously translates as "prism-structured eggshell from Colorado." Just to keep us on our toes, the name would later be changed to *Preprismatoolithus coloradensis*. Fear not, an explanation follows very soon.[3] But first let us rejoice that this eggshell type would, as noted below, be the basis of an egg family for which Karl can, according to the formal rules of scientific naming, forever take credit.

Karl would get into something of a skirmish, or minor egg war, with Bob Young. Bob was a well-known geologist in western Colorado who had written a few seminal papers on the sedimentary rock strata sequences, the field known as "stratigraphy," as well as a small book on local geology and wildflowers. Having found the site, he was naturally interested in the significance of the eggs, and he had the geological background to consider writing scientifically about the site, if not about the finer details of the eggshell.

In order to collect such valuable fossil eggs—some of which were quite complete, though crushed or flattened, which was not surprising after 150 million years of burial—one needed a permit, and a place to keep them in perpetuity. This meant that Bob and Karl would cross paths at the Museum of Western Colorado, the approved local repository. At that time, the museum's paleontology curator was Harley Armstrong, who would later become the state paleontologist, with permit authority for the federal public land management agency known as the Bureau of Land Management (BLM). He had first taken on his curatorial role at a time when the BLM was tightening up its fossil collection and management policies, not least because of a perceived need to combat certain rather egregious activity that one might characterize as very much in the Wild West outlaw tradition. It is beyond the scope of this present narrative to go into details or even to indirectly suggest the identity of any of the actors, but suffice it to say guns, damage to fossils, conspiracy rumors, and mental health had been involved. None of these historical shenanigans directly affected Bob, Karl, or Harley, although Harley played a significant role in mediating between what might politely be described as disagreement, if not a "turf war," between Bob and Karl.

It perhaps did not help that Bob's geological aspirations and background were more general than Karl's very specialized interests in eggshell structure. Bob was in fact using sieves to extract all the microfossils he could find from the site and was not focused on hard-to-find eggshell alone. There was also the technical question of how to extract the eggs. If there is more than one isolated egg, one does not just go in and hammer and chisel out individual eggs, or might we say "ovulates," in fist-sized chunks of matrix. Where several eggs occur, one might be dealing with a nest, even a nesting ground, and it could be important to excavate larger chunks of matrix so as to determine how the eggs were distributed. To do this, one has to excavate around a larger chunk of matrix and put it in a plaster jacket, a favorite technique of paleontologists, for wholesale removal. A crude analogy might be that at the supermarket one does not pick out one egg from a carton; one goes for a whole clutch in a carton or a crate.

In a joint 1987 paper, with Karl, Bob, and Harley literally on the same page and in this pecking order, the nest interpretation is explicitly laid out: "Similar eggshells restricted to a somewhat oval area, and the large number of eggshell fragments, representing more than one egg, suggest that this was a nest site, in which hatchlings reduced all eggshell to mostly small fragments. [4]

This seems reasonable enough and confirms that oval eggs may occur in oval nests, and Bob would later write a paper on it explicitly titled "A Dinosaur Nest in the Jurassic Morrison Formation, Western Colorado." One can imagine the intricacies, logistics, and potentially divergent opinions surrounding how best to extract the eggs in some sort of organized, geometrically oval arrangement so as to maintain the integrity of what might be remnants of a clutch or a nest. It's the proverbial strategy of putting all of one's eggs in the same basket. Differences of opinion were not smoothed over by the fact that Bob had home turf advantage and that he saw Karl as something of an impatient interloper—and as one who had fought on the "other" side in the not-yet-forgotten world war. Still, Harley was the consummate diplomat and was able to mediate between the warring factions; and, as we have seen, in the final analysis the eggs found a safe home in the museum and although not named after Bob Young, the label *Prismatoolithus coloradensis*, applied in 1994 and later changed to *Preprismatoolithus coloradensis*, can be seen as evidence that Colorado at least was neutral territory and had been important to egg-laying dinosaurs in the Jurassic.

But why, you may ask, while we are on the subject of names, does the "prismato" prefix bit become "preprismato," especially as the "pre" bit comes later? Well, it has to do with the rules of formal naming. When Karl coined the *Prismatoolithus* name in 1994, he did not know that some Chinese egg researchers (we'll meet them later) had proposed the same name (*Prismatoolithus*) the year before, in 1993. Both Karl and a couple of Canadian researchers had spotted the duplicate name and knew the Chinese name had historical priority. In fact, Karl suggested the Canadians propose the *Preprismatoolithus* label, which was published in 1996.[5] The moral of this shell game story is that you do not have to find or be the first to study a new specimen to coin a new name. In this case the Canadians, one of whom had worked closely with Karl, were simply providing an alternative name for the specimen Karl was soon to learn he had named incorrectly. If this proves to be the only mistake Karl made in the scientific name game, it amounted to a very minor technical error that only a few specialists even noticed. In any case, he was instrumental in correcting the error.

It was perhaps mildly regrettable that Karl and Bob Young had not seen eye to eye. Under different circumstances things might have been different. Perhaps over a beer. But you win some, you lose some. Karl enjoyed his beer, brandy, pipe, and cigars. There was and almost always is a certain social conviviality enjoyed when sharing a drink and a smoke, whether in a foxhole, around a campfire, or socializing with friends. Karl had low blood pressure. Thus, his doctor told him it was OK to drink and smoke, or so he told us with a twinkle in his eye! Doctor's orders. He mostly indulged in moderation, but true to his pledge to enjoy a new and better life, he knew how to loosen up when the spirits moved him.

Paleontologist David Gillette, who in 1986 had organized a Dinosaur symposium that Karl would attend, recalled another symposium story from a meeting in Philadelphia.[6] Ornithologist Pierce Brodkorb was arriving at the airport as hundreds of people converged and began to identify fellow attendees, even though they may never previously have set eyes on one another. All wandered toward the terminal exit. Just as birds of a feather flock together, so paleontologists tend to recognize one another's colors, most likely scruffy and weather-beaten from extended expeditions.

Converging on the taxi ranks and ready to flock together to share rides to their destination, Karl, Pierce, and Dave found themselves breaking the ice during a cab ride. With the next order of business being to set up base camp at the hotel, a common sense of purpose and destiny was soon consolidated with a convivial drink or two and a toast to the noble arts of paleontological investigation and brotherhood. Pierce had come prepared, bottle in bag. One toast led to another and another. Delving into history and prehistory, it soon became clear that both Karl and Pierce were World War II veterans and had fought on opposite sides on some of the same battlefields, perhaps even shooting at each other. But no hard feelings. The liquid refreshment, which Pierce had made sure to bring, was the magic elixir, ensuring truth, reconciliation, and the rapid and efficient execution of a peace treaty between the Allied and Axis powers, with no injury beyond the neutralization of a few brain cells, retreating in equal numbers on both sides.

Some paleontologists have grown up devoted to the local fossils, perhaps specializing in a narrow group at the recommendation of a graduate thesis advisor. Not Karl. Perhaps quite unwittingly—due to the nature of the subject of fossil eggs, which had been laid by all manner of species, all over the world—the world became Karl's scientific playground. Whether looking at eggs from a rare and sluggish New Zealand lizard look-alike or looking at condor-sized eggs, likely laid by *Allosaurus*, he hardly knew where the next discovery or egg hunt might lead.

In 1958, news of a couple of clutches of eggs had surfaced in the Canary Islands, off what had been the infamous, slave-trading Barbary coast of northwest Africa. A generation later it was time for Karl to get in the game and confirm that these rounded eggs, a little over two inches in diameter, were laid by large turtles or tortoises. Armed with his knowledge of the difference between the soft pliable shells of sea turtle eggs and the rigid eggshell of tortoises, Karl easily showed that they had been laid by tortoises. So he could show that these eggs, although found on islands, were "decisively not those of sea turtles."[7] Nor were they found in a sandy beach deposit. On the contrary, they were found in a well-known type of volcanic sediment known as a pyroclastic flow, caused by the eruption and lethal downslope avalanche of burning cinder- and ashlike sediment, which the French call a *nuée ardente* or "glowing cloud." It is a strange coincidence that the rocks that make up this deposit are called the Roque Nublo Group, literally the "cloud rock." The eggs, wherever they may originally have been situated, had evidently survived the conflagration. Had they been engulfed, buried, and fried instantly, or before burial? Had they been rolled or floated along on the glowing cushion of slurry like so much flotsam riding a hot avalanche? If they had been heated beyond a certain point, the eggshell would have shown signs of damage from the intense heat. Perhaps, as paleontologist Ken Carpenter suggested, "Very large land tortoises had dug a nest in the old volcanic debris."[8]

Instant, or at least quick, burial is a good way to be preserved for posterity as a fossil. The geological evidence shows volcanic eruptions dating from about four million years ago, and Karl's study of the eggshell shows they survived remarkably well. The egg layers were likely similar to the Galápagos giant tortoise, the Aldabra giant tortoise from the Seychelles Islands in the Indian Ocean, or fossil species from the African mainland. Although the Canary Islands are not that old, geologically, perhaps dating back ten million years, they have also produced giant fossil tortoise remains and various older eggshells, probably of ostrich ancestors or relatives. Unlike the four-million-year-old tortoise eggs, and an even younger one-million-year-old lizard caught up in a later glowing avalanche, the birds, whose eggshells are found in nonvolcanic sediments, had been spared a searing fate. One wonders about the fate of tortoises, famously slow-moving in the face of anything fast-moving. They would have stood little chance when facing an incandescent avalanche moving at up to fifty miles per hour and reaching a searing 1000°C. Will they too be found one day buried and baked in volcanic sediment? It is remarkable that such apparently fragile structures as eggs survived visibly intact when situated where they might be fried, baked, or broiled at the foot of an angry and unforgiving volcano.

Remembering that instant burial is a good way to be preserved for posterity, let us turn to a more recent volcano. When Mount Saint Helens erupted in 1980, unsuspecting gulls found their nests buried in volcanic ash.[9] In comparison with the incandescent glowing-cloud avalanches that roared down the volcano's slopes, further away ash fell more gently, without being superheated. Such ash-fall, comprising a mist of glassy shards, can still be lethal, especially if inhaled. Even two hundred miles away, the gulls' eggs were soon buried and suffocated. When the rains came, acid leached out of the ash and, as Karl and his colleague Jim Hayward determined, began to dissolve the eggshell, leaving only a ghost of its original structure. Within a year the eggshell became pitted, and progressively it dissolved and degraded. Strangely, as Karl and his colleagues pointed out, the Canary tortoise eggs had survived in extreme volcanic terrain for four million years of geological history; yet despite having rigid shells the gulls' eggs were not destined to be preserved as elite fossils. Each situation is different, and paleontologists need to study not only what is preserved but also how it is preserved.

Not every rock or fossil is easily identified. The University of Colorado Museum, where Karl's collection would eventually find its permanent home, has a large collection of trace fossils, as well as the usual assortment of body fossils. The trace fossils can be broadly divided into two categories: footprints (or tracks) and coprolites (poop or feces); eggs are a third, rather special category. Paleontologists recognize nests as trace fossils rather like footprints, but eggs are actual biological entities, though obviously different from the fossilized body parts of the egg layers. Oddities or pseudofossils in all these categories sometimes pop up as curiosities brought to museums by amateur collectors, or even by puzzled professionals. Rock slabs with indentations may look like footprints but just be holes in the rock. Lumpy nodules may look like petrified poop, and rounded pebbles may look like eggs. A common mistake is to assume that some dinosaur eggs were the size of beach balls, a guaranteed strain on a mother's oviduct, when in fact none are known to be any larger than an elongate American football, or rounded soccer ball, and many are much smaller, orange- or grapefruit-sized.

On one occasion Karl received a very regular-shaped nodule, about two and a half inches long, which he described as "outwardly . . . much like a fossil egg," and about the size of one laid by today's hens. But don't judge a book by its cover. "When the specimen was X-rayed," the core was revealed to be a lead fishing sinker, surrounded, like layers of an onion, by six layers of the chalky mineral brushite. It turns out the specimen was a "stomach stone" or "calculus," also a name for kidney stones and gallstones, and nothing to do with mathematics.[10] They can form in the stomach of

When an egg is not an egg, it could be a "madstone," a pseudo-egg created when a large ruminant animal swallows a foreign object. In this case, a fisherman's lead sinker was swallowed and coated with a chalky mineral layer of calcium phosphate known as brushite.

hoofed, cud-chewing animals or ruminants such as deer, buffalo, and sheep, renowned for their powerful digestive systems. If such animals swallow foreign objects, like little pebbles, or even fishing sinkers, their digestive systems begin to coat these unwanted and irritating foreign objects with protective mineral layers that may help reduce serious indigestion. Such stomach stones are quite different from the gizzard grit that birds and some dinosaurs deliberately swallowed to help grind up their food because they lacked the digestive power of cud-chewing mammals. These polished gastroliths, sometimes called dinosaur "belly boulders," and sometimes tennis-ball-sized, are polished to a shine by dinosaurian stomach acid. Such curiosities have also been the subject of a few specialized studies, especially when found en masse in the stomach position in complete skeletons.

Examples of the onion-layered calculus have been known since at least the fifteenth century, when they were called "madstones" and thought to have had healing properties. Intrigued by the madstone phenomenon, Karl examined at least ten specimens, found at various sites in Colorado, Utah, Wyoming, Arizona, and North Carolina, which he described as having a "loose crystal structure" and "texture similar to that of a reptilian eggshell."[11] Coupled with the ovoid shape, it is no surprise that they could be taken for eggs. Their organic origins were often given away by the "putrid odor" they gave off when cut open. In case you're wondering, they were mostly all made of chalky brushite, or calcium phosphate, a mineral found in bone, teeth, and even milk.

Karl had answered the question "When is an egg not an egg?" Trackers and poop experts have to answer similar questions. When is it not a track, but just a hole in the ground? And when is pseudo poop just pseudo poop? If you're asked such questions, you may be, like Karl, a presumed 'trace fossil' expert, or in the case of stray ovoids, an eggs expert.

Notes

1. After reading a paper to the Geological Society (London) in 1859, John Buckman published the paper, in which he described *Olithes bathonicae* from England. It was the first fossil egg to be scientifically described. See J. Buckman, "On Some Fossil Reptilian Eggs from the Great Oolite of Cirencester," *Quarterly Journal of the Geological Society of London* 16 (1860): 107–110.

2. K. F. Hirsch, "Upper Jurassic Eggshells from the Western Interior of North America," in K. Carpenter, K. F. Hirsch, and J. R. Horner (eds.), *Dinosaur Eggs and Babies* (Cambridge: Cambridge University Press, 1994), 137–150.

3. D. K. Zelenitsky and L. V. Hills, "An Egg Clutch of *Prismatoolithus levis* oosp. nov. from the Oldman Formation (Upper Cretaceous), Devil's Coulee, Southern Alberta," *Canadian Journal of Earth Science* 33, no. 8 (1996): 1127–1131. It is of interest to note that Karl was in touch with Zelenitsky and Chinese researcher Zhao and, as he was aware of the need to propose a new name, advised Zelenitsky to do so (Zelenitsky, written communication to the authors).

4. K. F. Hirsch, R. Young, and H. J. Armstrong, "Eggshell Fragments from the Jurassic Morrison Formation of Colorado," in *Paleontology and Geology of the Dinosaur Triangle, Guidebook,* ed. W. Averett (Grand Junction: Museum of Western Colorado, 1987), chap. 12, n.11.

5. Initially, unbeknownst to Karl, the Chinese researchers Zhao and Li (cited below) had proposed the name *Prismatolithus* for Chinese eggs a year before Karl proposed the same name for his Colorado specimen (Hirsch, "Upper Jurassic

Eggshells from the Western Interior"). Therefore, priority rules made it necessary to provide a new (*Preprismatolithus*) name for Karl's specimen, as done by Zelenitsky and Hill (Zelenitsky, "An Egg Clutch of *Prismatoolithus levis*"). For Karl this was not a surprise relabeling of his specimen: he in fact suggested the revised naming to his Canadian colleagues, as verified in the acknowledgment of the Canadian paper. Z. Zhao and R. Li, "First Record of Late Cretaceous hypsilophodontid Eggs from Bayan Manduhu, Inner Mongolia, *Vertebrata Palasiatica* 26 (1993): 107–115.

6. See chapter 15.

7. K. F. Hirsch and L. F. LopezJurado, "Pliocene Chelonian Fossil Eggs from Gran Canaria, Canary Islands," *Journal of Vertebrate Paleontology* 7, no. 1 (1987): 96–99; see page 98.

8. K. Carpenter, *Eggs, Nests, and Baby Dinosaurs* (Bloomington: Indiana University Press, 1999), 112.

9. J. L. Hayward, K. F. Hirsch, and T. C. Robertson, "Rapid Dissolution of Avian Eggshells Buried by Mount St. Helens Ash," *Palaios* 6 (1991): 174–178.

10. Karl wrote two short articles about these curious "madstones." For the article for the popular press, see K. F. Hirsch and H. L. Robinette, "Looking into a Madstone," *Colorado Outdoors* 35, no. 2 (1986): 23. See note 11 for a second, technical article.

11. In K. F. Hirsch, "Not Every "Egg" Is an Egg, *Journal of Vertebrate Paleontology* 6 (1986): 200–201, see page 200.

CHAPTER FOURTEEN

From the Gobi Desert to
Egg Mountain

No occupation is more worthy of an intelligent and
enlightened mind that the study of nature and natural objects. . . .
We shall constantly find some new object to attract our attention,
some fresh beauties to excite our imagination, and some previously
undiscovered source of gratification and delight.

Sir John Paxton, 1838[1]

Before we step into Cretaceous egg territory, it was historically necessary to tell the aforementioned (chap. 13) stories of the Wild West dinosaur discoveries that allowed Colorado and neighboring Wyoming and Utah to claim themselves to be the original Jurassic Park, a series of fossil fields abundantly populated with iconic Jurassic dinosaurs like *Diplodocus*, *Stegosaurus*, and *Allosaurus* and the scarce remains of a few inconspicuous eggs and nests.

Before this large swath of Wild West fossil-hunting territory became well known in the 1870s and 1880s, post-Jurassic dinosaurs from the Cretaceous period were known in modest numbers from here and there'round the world. In fact, *Iguanodon*, the second dinosaur ever described from England, in 1825, was Cretaceous in age and the ancestor of well-known duck-billed species that would be found abundantly in North America and Asia.[2] However, it was not until the early twentieth century that these Cretaceous fossil fields opened up in the Wild West US and western Canada, as well as in China and Mongolia. To this day, the Cretaceous deposits of North America and Asia and some other regions still play a leading role in the field of dinosaur eggs (oology), nests, nesting sites, babies, and reproduction.

No history of dinosaur paleontology is complete without the story of American Museum of Natural History (New York) expeditions to Mongolia in the 1920s.[3] Not only were the expeditions a well-publicized success, they also spurred other nations to head for the Gobi in search of fossil treasures. Despite the American discoveries of entirely new dinosaurs such as *Protoceratops* (a pig-sized ancestor of more famous horned herbivorous dinosaurs like *Triceratops*) and the equally famous *Velociraptor* (meaning "speedy predator"), the expedition became most famous for the first discovery of large numbers of dinosaur eggs, preserved as photogenic clutches, assumed to occur in nests, which seemed to lack any visible structure to distinguish them from the surrounding sediment. For some reason the public often found the image of nest-making, egg-laying, and presumably egg-brooding dinosaurs almost more captivating than the dinosaurs themselves.

Another feature of this expedition that garnered huge attention was the swashbuckling figure of the American Museum paleontologist Roy Chapman Andrews, a larger-than-life figure when he led the expedition, cowboy-style, into the Gobi Desert with his bandit-repelling pistol on his hip and an explorer's outfit that two generations later would make him the model for the movie hero Indiana Jones. Capitalizing on his larger-than-life persona, Andrews would write many popular accounts of his travels in Asia where, as well as dinosaurs, he hunted rare game and would do the taxidermy himself. He was also an expert on living and fossil whales. As well as his scientific papers and expedition reports, he also wrote popular children's and junior books for aspiring paleontologists.

By the time Karl was a toddler, finding his infant feet in 1922 Berlin, the thirty-eight-year-old Andrews had already been biologizing around southern China and had visited Beijing (then Peking) and Mongolia, making preparations for the Gobi expeditions that would be called the Central Asiatic Expeditions. This ambitious international project ran from 1922 to 1928, ending when Karl was turning seven. For good measure, and a splash of historical color, the horse Andrews first rode in Mongolia was christened Kublai Khan, in honor of the first emperor of China's Yuan Dynasty. Kublai Khan was grandson to the notorious, steppe-riding, nomad emperor Genghis Khan, who had razed and terrified civilizations as far west as what would become Karl's native Germany.

Andrews's Mongolian haul of paleontological treasures contained so many dinosaur fossils and other fossils, especially those of small mammals, that little serious attention was paid to the dinosaur eggs. Andrews even promulgated stories about the egg layers that would later prove wrong or at best speculative. It was left to those on the Eurasian continent, where Karl had his cultural roots, to study the Mongolian eggs in the detail they deserved. In fairness, nobody was studying fossil eggs seriously in the 1920s. The eggs were what Victorian naturalists liked to call natural "curiosities"—books judged largely by their covers, with their inner secrets and origins still mysterious. It would take another two generations before Karl and others got serious about the dinosaur egg business.

A native of Montana's big sky country, Jack Horner[4] is one of North America's colorful and better-known dinosaur hunters, sharing a memorable name with little Jack, the protagonist of an eighteenth-century nursery rhyme. In the pioneering spirit of the Wild West, Jack was a self-made man who came to paleontological prominence in the 1970s as a result of some groundbreaking discoveries. Unlike pie-eating, corner-sitting little Jack from the nursery rhyme, who "stuck in his thumb, and pulled out a plum," Montana Jack stuck both thumbs into the dirt of the badlands, at what would later be called "Egg Mountain," a rather small badlands hummock, and pulled out dinosaur eggs, nests, and babies.

Jack and Karl would become friends and scientific colleagues, eventually publishing together. Although Karl was a generation older, they shared various common traits and interests besides fossils. Both were war veterans who loved the wide-open spaces. Jack had reputedly spent his early months living in a tent while his father started up a gravel business. Karl too would camp with Jack in the field, at Egg Mountain, famously sheltering in teepees, well-suited to the windswept plains that Crow and Blackfoot tribes had once called home.

Jack's fame developed as a result of two convergent circumstances. One was his natural curiosity about fossils, and the other was the accident of birth that had put him in Montana's "Big Sky Country," which paleontologists might equally well call "Big Earth Country," where few had scratched the prairie surface deep enough to bring more than a glimpse of prehistory to light. Jack, like others of his generation, had come of age during one of dinosaur paleontology's periodic upswings, which historians in the field like to call "golden ages." A century after the 1870s' opening up of the West's dinosaur-digging fields, which brought many of the most iconic Jurassic and Cretaceous dinosaurs to scientific and public attention, a new "dinosaur renaissance" was underway. This new wave was infused with a novel ecological consciousness that saw dinosaurs as more than giant, big-game trophies, instead regarding them more respectfully as dynamic and once-living species within ancient and complex ecosystems. The old stereotype of extinct lumbering giants, which implied failure due to lack of adaptable intelligence, no longer satisfied the new generation of paleontologists. By the 1970s, new questions were being asked. Principal among them: How come dinosaurs survived and were successful for so long? Were they warm-blooded like today's birds and mammals? Did they have complex social and family lives? How did they reproduce? Where did they nest, lay eggs, and rear their young? How fast did they grow? There were many such questions, and whether he knew it or not, Jack was on the verge of finding some significant answers, and opening new fields of inquiry in dinosaur paleobiology, into which Karl would be a natural and timely fit.

Jack's most notable discoveries from Egg Mountain are now a well-known chapter in the history of American paleontology. What he found was that duck-billed dinosaurs had nested in large numbers and had also returned to their nest sites over many seasons. Evidently they were colonial nesters, like some modern birds, and faithful to their preferred site. Jack called it "site fidelity," the first example known among dinosaurs. He also showed that some nests were full of the bones of baby nestlings that had grown up to about three feet long, far bigger than the size of hatchlings, just out of the egg. This indicated that the babies had stayed in the nest for some time, probably at least a few months. This in turn meant the parents must have cared for them. Thus, Jack named the dinosaur *Maiasaura*, meaning "good mother" dinosaur. Maia was a Greek mother goddess. It was also the first time a dinosaur had been given a name with the female ending "saura" rather than the male ending "saurus." This, and some may smile wryly, was one of Jack's contributions to 1970s gender equality in the world of vertebrate paleontology.[5]

Karl, as an egg expert, knew immediately that Egg Mountain was significant for another reason. Here were eggs and eggshells that could be matched to known dinosaurs. As a bonus, Jack had found evidence that other dino-saurs had nested here. Jack initially attributed the non-*Maiasaura* eggs to a small duck-billed relative, known as *Orodromeus makelai*, that appeared to have been able to leave the nest soon after hatching. Such dinosaurs are called precocial (meaning "precocious"), in contrast to altricial species, like *Maiasaura*, that developed more slowly. It was later shown that *Orodromeus* was not the egg layer. Nevertheless, the name is somewhat iconic in opening up debate about the difference between altricial and precocial development, or what we may call different "reproductive strategies." *Orodromeus makelai* (meaning "mountain runner"—that is, on Egg Mountain) had been named by Jack after Bob Makela, his long-time friend and field crew chief, whom Jack first met when Bob came into a university class carrying a poisonous Gila monster.[6] *Orodromeus* is a valid dinosaurian name, but further study of the embryonic remains found in the supposed *Orodromeous* eggs proved them to be those of the theropod dinosaur *Troodon*.[7]

Bob and Karl shared a birthday on March 20, the spring equinox, and together would celebrate as drinking buddies after a day's field work. Bob's skill as a field crew chief involved getting beer and other supplies into remote camps. Both Bob and Karl were destined to play important roles, as the an-nals of paleontology already show, thanks to the various lines of evidence they uncovered surrounding the birthing of dinosaurs and the birth of a new chapter in the story of how dinosaurs came into the world. If, like birds, dinosaurs gave birth to the next generation in the spring, there may have been some that laid their eggs, incubated them, and perhaps hatched them in the spring equinox season, when today's paleontologists aspire to get out in the field to find new evidence of ancient life.

Besides their common inclinations for fossil-hunting, the freedom of wide-open spaces, and periodic liquid refreshment in field camp, Karl and Jack shared something else. They both made their debut in scientific publishing in the late 1970s. Neither had had the opportunity to shine academically early in life, but that would not stop them now. Both had found ways to fa-miliarize themselves with the academic world so as to learn the norms and conventions of scientific research and publication. With the help of Judith Harris, at the University of Colorado, Karl had put together his first scientific paper, "The Oldest Vertebrate Egg?," published in 1979.[8] Jack, who through his frank autobiography, *Digging Dinosaurs*, would later reveal his struggles with dyslexia and an aversion to traditional schooling, would publish his first paper with friend Bob Makela, in 1979, in the prestigious journal *Nature*, under the title "Nest of Juveniles Provides Evidence of Family Structure Among Dinosaurs."[9] In many ways this paper helped launch Jack's paleontological career. By the end of the 1980s, Jack had published two other papers in *Nature*, one in *Scientific American*, and his popular autobiographical book *Digging Dinosaurs*.[10] Karl, meanwhile, by the end of this same decade had racked up twenty-five papers on eggs and eggshells and had been a regular visitor to the Museum of the Rockies, in Bozeman, Montana, where Jack was now a celebrated and increasingly well-funded paleontologist. As Karl would put it, in 1990, Jack's discoveries of "nesting areas with clutches, nests with hatchlings, eggs with embryonic remains . . . had great impact on the study of dinosaurs and led to the discovery of more fossils eggs and eggshells."[11] However, despite these important finds, at sites that became known as Egg Mountain, Egg Island, and Egg Gulch, Karl would note, in 1990, that " there is still little known about the structure of the eggshells." Jack no doubt deserved the fame and approbation that came with finding cute baby dinosaurs, fed by loving mothers. He had earned the rewards of his hours scouring the badlands for fossils, but there was still much to be learned about the egg, the development of the cute hatchlings, and the behavior of the colonial nesters. Jack would recall thinking that it might be educational to wander into a gull colony to get a sense of how these dinosaur descendants organized their seasonal colonial nesting. He got more than he bargained for. An ornithologist might have warned Jack that he would face a reception committee of screaming and extremely irate gulls. Angry birds have no hesitation in mobbing invaders, swooping down on their heads with raucous screams, and even emptying rank-smelling bowel contents onto any unwelcome visitors.

Did dinosaurs in the Cretaceous exhibit the same parental instincts? Among the 1920s discoveries of Andrews's team in Mongolia was a find that suggested this may have been the case. The discovery consisted of the skeleton of a carnivorous theropod dinosaur named *Oviraptor*, preserved with its head just a few inches above a nest that was thought to represent that of the smallish pig-sized vegetarian dinosaur *Protoceratops andrewsi*. Put evidence of a carnivore and a vegetarian dinosaur so close together, and a predator-prey scenario is too much to resist. So it looked, at first sight, as if the *Protoceratops* was victim of theropod predation, as the name *Oviraptor*, meaning "egg thief," implies.[12] *Oviraptor* was actually a beaked, toothless dinosaur, perhaps a swallower of eggs but not a slasher of flesh. As we shall see, however, the label "egg thief," now enshrined in the paleontological literature, is a slanderous misrepresentation that took generations for *Oviraptor* to shake off. The slander was indirectly reinforced almost fifty years later when another theropod, *Velociraptor*, was found locked in an apparent to-the-death struggle with a *Protoceratops* in the same region of Mongolia. *Velociraptor* had nasty serrated teeth, reminding paleontologists of the proverbial flesh-slashing steak knife, and for good measure it had a nasty slashing sickle claw, common to this group. This makes it, like the egg thief, a member of the raptor tribe, which only added to its notoriety, as new and larger raptor species were discovered—and incorporated, as villains, into such movies as *Jurassic Park*.

However, although vindication of *Oviraptor* took more than seventy years, much more recent American Museum expeditions to Mongolia have finally confirmed that *Oviraptor* was in fact another good parental dinosaur. Possibly the mother, but it could equally well have been the father or both parents, brooded the clutch.[13] Although "she" was the egg layer, it is not proven that she was the one found in brooding position on nests, and the one that stayed with the clutch until buried and killed under catastrophic inundations of sediment.[14] As recent commentators would later point out, *Oviraptor* was falsely accused of robbery and would also have been successful in pursuing an international paternity suit against Andrews's American team for mistakenly attributing its eggs to an entirely different species (*Protoceratops andrewsi*).

Ironically, it was the same fierce parental instincts that Jack encountered in the gull colony that had made Cretaceous dinosaurs such good dinosaurian parents, whether of Mongolian or Montanan pedigree. Jack had helped make dinosaur nest and egg clutches from Montana famous, by finding the evidence that allowed science writers to tell heart-warming stories of dedicated good-mother dinosaurs tending to cute babies. Thus, in both Mongolia and Montana the stories carried subtexts about the bravery of dinosaurian parents valiantly defending their nest from predators, because in dinosaur paleontology it is always safe to assume there is a rapacious predator lurking nearby, whether one finds direct evidence or not. It is judicious to remind ourselves of the uncertainty about which dinosaurs even guarded or brooded their nests rather than burying eggs like crocodiles or turtles. Perhaps, like living birds and mammals, dinosaurs had a wide range of parental care behaviors.

But there is more to egg research than pictures of nests and reconstructions of lovable babies and protective parents, however appealing such images are to kindergarteners and the cartoon industry. All these newsworthy reports on eggs and nests were still about external appearances, those books with fine covers but mostly blank or very sketchy pages inside. The challenge was to get "inside" the egg, its eggshell structure and secrets. It was in the countries of the Old World that many fossil egg secrets would be cracked open. This would give Karl the opportunity to reconnect with discombobulating, not-always-comforting, but perhaps psychologically reconciliating memories from earlier chapters in his life.

Underpinning the still-embryonic study of fossil eggs in the early 1980s, when Karl entered the field, were a series of studies by unassuming scholars who were interested, like Karl, in the structure of eggshell. Fortuitously, some of the seminal studies had been undertaken by Germans, writing in German. For example, Wilhelm Nathusius, whom Konstantin Mikhailov describes as a "genius," recognized that eggshell is not merely a mineral, crystalline structure.[15] Rather it is, as noted previously, a complex network of organic and crystalline "tissue," best described as a composite or biocrystalline—that is, a biological entity, not a "trace fossil."

As noted in chapter 9, Karl would familiarize himself with the work of Professor Heinrich Erben, at the University of Bonne, who had pioneered the field of "biomineralization" in the late 1960s and was editor of a German journal with the same name. Erben had begun to compare the structure (technically the microstructure) of turtle, gecko, crocodile, bird, and dinosaur eggshell, and Konstantin (also an egg expert, and a Russian, who became Karl's close friend) had advised Karl to read this work. Perhaps, initially thinking of Karl as a keen amateur, Konstantin would write this:

> Thus, when our correspondence started, I . . . tried to explain [to] Karl some things that he did not understand in the writings of Erben. And advised him to read this and that. *And he did!* He was absolutely . . . without any empty ambition [to] excel in our cooperation. Accepting the fact of my serious education in biology (and I always shared my knowledge with great pleasure), . . . I could only wonder how at his age he was willing to learn more and more and did it with such a pleasure![16]

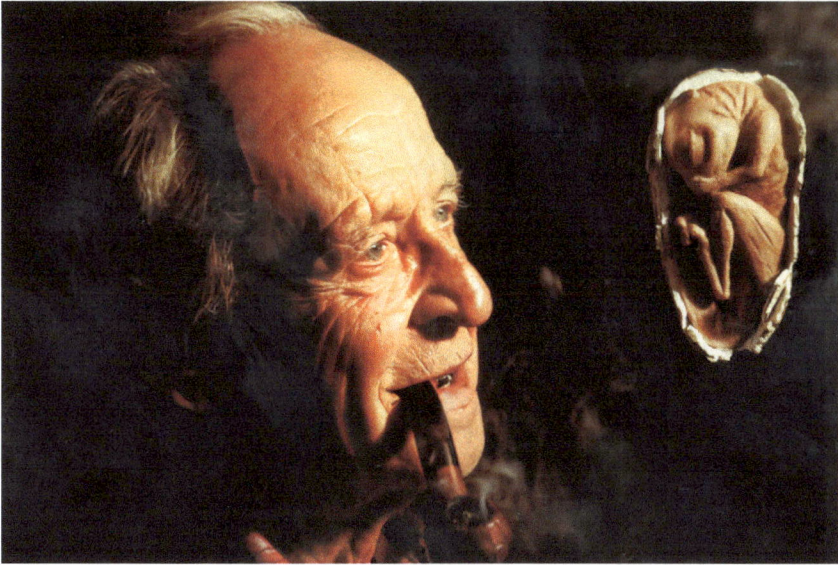

An iconic photo of Karl with pipe, egg, and embryo, taken by Louis Psihoyos,
a former *National Geographic* photographer and the author of *Hunting Dinosaurs*.
Used with permission (see note 17).

International science was proving a great leveler. For all the disadvantages
faced by those whose education is deficient in some way, such as the inability
to read German, an important scientific language (and whose education is
not lacking in some way?), lack of fluency in German was no impediment to
Karl. In these early days of the 1970s and 1980s, Karl was embarking on the
ramp to the so-called "learning curve" and would take his time to absorb
what he needed to know. Whether consciously or not, he had been laying a
solid foundation for the heady days of the 1990s, when he would be hoisted
shoulder-high as a celebrated "dinosaur egg man," hanging out with Jack
Horner and other superstars of dinosaur paleontology. He would be featured
as "The Egg Man" by Louis Psihoyos, then a *National Geographic* photographer, in his 1994 book *Hunting Dinosaurs*. The book was an elaborate spin-
off of a 1993 article, for which Psihoyos had traveled the world for eighteen
months taking forty thousand photographs, of which only about forty stun-
ning studies made the editor's cut. Karl is quoted as saying modestly that his

twenty years of study were a mere "hobby" but one that "hooked" him. He was, he claimed, "the *worst* authority because I am the only one."[17]

One of the dubious pleasures enjoyed by the world's best, worst, or only egg expert was the occasional receipt of bizarre correspondence. One gentleman claiming to own a fossil egg insured for $1 million by Lloyds of London had not only written to Karl with his claim but placed him in elite company by also forwarding the claim to the Queen of England, the Prince of Wales, Pope John II, and the famous American talk show host Johnny Carson.

Notes

1. This quote by Sir John Paxton comes from his book *A Practical Treatise on the Cultivation of the Dahlia*. https://todayinsci.com/QuotationsCategories/E_Cat /Enlightenment-Quotations.htm

2. *Iguanodon*, meaning "iguana tooth," was the second formally named dinosaur, reported in 1825 from Cretaceous rocks in England (*Megalosaurus*, named in 1824, was the first). At the time, the concept of "dinosaurs" did not exist. The group name Dinosauria was proposed in 1841 and published in 1842. At that time only three incomplete dinosaur skeletons were known, all from England.

3. The American Museum expeditions to Mongolia (1921–1930) are among the most famous in the annals of paleontology. They were first reported in H. F. Osborn, "Ancient Fauna of Mongolia Discovered by the Third Asiatic Expedition of the American Museum of Natural History," *Science*, 57, no 1487 (1923): 729–732. A complete 678-page report, entitled *The New Conquest of Central Asia: A Narrative of the Explorations of the Central Asiatic Expeditions in Mongolia and China, 1921–1930*, was published by the American Museum in 1932, with Roy Chapman Andrews as the senior author. Andrews was a dashing, adventurous character, and by far the best known of the paleontologists to participate in these expeditions. In addition to many scientific papers, he published at least two dozen popular books, including several that dealt with his Chinese and Mongolian adventures. In his honor, his colleagues named one of the best-known Mongolian dinosaurs *Protoceratops andrewsi* (see notes 11 and 12 and associated text).

4. Jack Horner became famous with the general public after the publication of his first book—J. Horner and J. Gorman, *Digging Dinosaurs: The Search That Unraveled the Mystery of Baby Dinosaurs* (New York: Workman Publishing, 1988)—in which he tells the story of his ground-breaking discoveries of Cretaceous dinosaur eggs, nests, babies, and parental care in Montana. See note 5.

5. J. R. Horner and R. Makela, "Nest of Juveniles Provides Evidence of Family Structure among Dinosaurs," *Nature* 282, no. 5736 (1979): 296–298.

6. J. Horner and D. Weishampel, "A Comparative Embryological Study of Two Ornithischian Dinosaurs," *Nature* (London), 332 (1988): 256–257.

7. J. Horner and D. Weishampel, A Comparative Embryological Study of Two Ornithischian Dinosaurs, *Nature* 383 (1996): 103. This paper is a "correction" of the interpretation that embryos from Egg Mountain were those of *Orodromeus* (note 6). They were shown to be those of the theropod dinosaur *Troodon*.

8. See chapter 10, note 4.

9. See note 5.

10. See notes 4–6, and R. Horner, "The Nesting Behavior of Dinosaurs," *Scientific American* 50 (1984): 130–137.

11. In K. F. Hirsch and B. Quinn, "Eggs and Eggshell Fragments from the Upper Cretaceous Two Medicine Formation of Montana," *Journal of Vertebrate Paleontology* 10, no. 4 (1990): 491–511, see page 491.

12. The supposed egg thief *Oviraptor* was described by H. F. Osborn, "Three New Theropoda, *Protoceratops* Zone, Central Mongolia," *American Museum Novitates* 144 (1924): 1–12.

13. There is little evidence that female dinosaurs were the primary egg brooders. This role may have been shared by females and males or could even have been mainly the province of males. J. R. Moore and D. J. Varricchio, "The Evolution of Diapsid Reproductive Strategy with Inferences about Extinct Taxa," *PLoS ONE* 11, no. 7 (2016): e0158496.

14. M. A. Norell et al., "A Nesting Dinosaur," *Nature* 378, no. 6559 (1995): 774–776.

15. Reminiscences of Konstantin Mikhailov sent to Martin Lockley (December 2018) via email. This 2,300-word document has been added to the UCM Karl Hirsch archive.

16. Reminiscences of Konstantin Mikhailov.

17. L. Psihoyos and J. Knoebber, *Hunting Dinosaurs* (New York: Random House, 1995), 297.

CHAPTER FIFTEEN

From Russia, and China, with Love

While Karl was still a prisoner of the Soviets in 1946, paleontologists from Moscow's renowned Paleontological Institute, under the flag of the Academy of Sciences of the USSR, were digging up their first dinosaurs in Mongolia as part of the Mongolian Paleontological Expedition (MPE), also known as the Soviet-Mongolian Paleontological Expeditions (SMPE).[1] The Soviet intention had been to launch the expedition in 1941, but after Germany attacked the USSR that June, the MPE was postponed. With what Soviets called the "Great Patriotic War" over, though not yet for Karl, the MPE could get underway, in August 1946, trundling 4,700 kilometers from Moscow to Mongolia's capital Ulaanbaatar. Despite seven months in the field, much of the 1946 expedition was a reconnaissance mission, and much of 1947 was spent in preparation for a second expedition, which would run for eleven months. Famously, this expedition would open up the so-called Dragon's Tomb locality, from which complete skeletons of the *T. rex*–like dinosaur *Tarbosaurus* and the duck-billed dinosaur *Saurolophus* were excavated. As noted by the Polish geologist Karol Sabath, the Soviet expeditions also "found eggs and eggshells, described later in Russian, by Andrey Sochova," one of several pioneers in eggshell classification.[2]

The SMPEs were led by one of the giants of Soviet paleontology: Ivan Antonovich Yefremov, the surname sometimes spelled "Efremov," who pioneered the field of taphonomy, the study of what happens to fossils between death and final burial. This is particularly useful in helping to explain why some skeletons are complete, with even their fossilized skin intact, whereas others are disarticulated—that is, broken up by erosion or scavengers. Eggs and eggshell are also subject to the processes of taphonomy, as we saw in earlier chapters dealing with hazards such as the dissolving of gulls' eggs. Efremov was something of a romantic, larger-than-life character, perhaps the Soviet equivalent of Roy Chapman Andrews. As a successful science fiction writer, he wrote several popular fictionalized accounts drawing on his expeditions to the Gobi Desert in Mongolia, with such evocative titles as *Road of Winds* and "Shadow of the Past."[3] The latter, which is available in English, includes a preface with insights into Efremov's scientific and literary philosophy. For example, extolling the virtues of the science fiction genre, he wrote, "My 'far away countries' are also new roads still to be trodden . . . I . . . try to lift the curtain of mystery over these roads, to speak of scientific achievements yet to come . . . as realities . . . to lead the reader to the most advanced outposts of science. . . . Its philosophy is to serve the development of the imaginative and creative faculty. . . . The subject of my writings will be scientific research, discovery, travel."[4]

By the time Efremov penned these sentiments in 1953, he had decades of field experience in Central Asia to inform a depth of paleontological knowledge. Karl, only thirteen years his junior, had spent barely a year on American soil, and his field adventures and study of Central Asian dinosaur eggs and Soviet paleontological literature lay ahead. However, Karl and Efremov had been cast from similar adventurous molds, and both were destined to share similar experiences treading dusty roads in search of rare fossil treasures:

> The vans swayed and bobbed . . . along the endless steppe . . . on all sides grey . . . and sun-scorched . . . some 250 miles away . . . from town—250 miles of . . . dunes, rocky hills, flat steppes . . . the waterless rocky desert in whose heart lay the graveyard of prehistoric monsters.[5]
>
> "Picture for a moment," said the palaeontologist, . . . "the endless chain of generations . . . that passed in those hundreds of millions of years. Try to imagine the unimaginable—the number of sacrifices placed on the altar of natural selection in the course of . . . evolution." An eagle cried from a dizzy height. . . . No one said a word. All eyes were fixed on the palaeontologist. He smiled thoughtfully and said, "Yes, friends, the greatness of palaeontology lies in its vast perspective of time."[6]

Although not fully appreciating their future role, the Soviets were, in the late 1940s, deep in dinosaur egg territory, which had already been skimmed over by the Americans in the 1920s and was already mostly forgotten for a generation. The Soviets were examining what Efremov would call "the riddle of the dinosaurian graveyards in Central Asia."[7] Little did Karl know—at that time still in his mid-twenties, and still a prisoner, as the expedition trucks

Illustration by G. Petrov of a typical paleontological expedition scene in Central Asia, used by Ivan Efremov to illustrate his story *The Shadow of the Past.*

rolled east from Moscow in the general direction of his internment camp—indeed, little could he know, what was in store. Soviet discoveries would lead, after another generation in which a liberated Karl would become a bona fide paleontologist, to a refined Soviet appreciation for the subtleties of the structure of fossil eggshell. Karl would eventually share, with a few specialist Soviet and German egg lovers, the mantle of pioneer in this field of study, and he would, as we have seen, make friends with Konstantin Mikhailov, who represented a link to the next generation of egg men and egg women.

Jump forward from the dark 1940s to the rejuvenating 1960s—with Karl safely on, and digging in, the soil of the American Wild West—and we find the so-called Dinosaur Renaissance underway. Between 1964 and 1971, a series of Polish-Mongolian Paleontological Expeditions followed in the footsteps of the Soviet expeditions of the late 1940s. As neighbors to Mongolia, the Soviets had never entirely left the Gobi, where they would serve as some-time advisors to the Mongolian government, keeping their hand in with the "diplomatic games connected with the People's Republic of China which had just been created" in 1949.[8] In this same 1960s and 1970s period, the Soviets ran various Soviet-Mongolian expeditions and discovered "huge clutches of sauropod eggs." These would prove to be the catalyst of fossil egg research that was needed to do justice to the many new finds of eggs, nests, and babies of dinosaurs and other vertebrate groups that were yielded up by the rich fossil fields in North America, across the world, but especially in Mongolia and China. Polish paleontologist Karol Sabath would describe the Soviet and Polish egg collections as among the "world's most diverse" and abundant, attracting not only American interest in the 1920s and Soviet-Mongolian and Polish-Mongolian interest through the 1970s, but since then additional joint Mongolian ventures with American, Canadian, Chinese, and Japanese teams.[9] Another review of the Russian-Mongolian expeditions referred to "19 forms of dinosaur eggshells."[10] This report was later fine-tuned to speak of "nine dinosaurian egg oofamilies containing twenty-one genera oogenera, and two avian oofamilies represented by three oogenera."[11] Thus, more like twenty-four different forms. We shall learn more of such methods of classification below. The reader might do well to remember that all these discoveries helped place the findings of skeletal remains in context and build up a picture of the paleoecology of likely nesters, egg layers, and their contemporaries.

Such an abundance of treasures had heralded a new chapter in dinosaur-related paleontology, while new ideas about dinosaur ecology and behavior were generating intellectual stimulation and enthusiasm. A further catalyst had been provided by new technology, including the electron microscope, which by the 1970s became a standard tool in the natural sciences and soon thereafter in Karl's eggshell studies. The fruits of these radical new studies were becoming conventional wisdom, or at least accepted paleontological wisdom.

By the 1980s, as Karl, Jack, and others hit their stride in the dinosaur-eggs-and-babies department, both in the field and squinting into microscopes, Russian egg expert Konstantin Mikhailov was emerging on the scene to become well known for his work on the systematic classification of eggshells. Karl would chuckle that a rising star in the field—one with whom he had corresponded and collaborated and who was, like Karl, a specialist pioneer—was a Soviet citizen (Karl always said "Russian") from the country that had held Karl captive and against which he had been sent to fight. But it was part of Karl's destiny to hold no grudge and even to look on his former prison guards as human beings who had, like him, been thrown in the maelstrom of war for no obvious or rational reason. Ironically, Karl, in his adopted America, was perhaps freer to pursue his scientific calling, and at least to travel widely, than those who until the 1990s were still behind the Iron Curtain and in its lingering shadow.

Humans have certain magpie qualities. They like to collect pretty objects. This can often develop into a collecting mania, as was rampant among many in the nineteenth-century heyday of natural history exploration and travel. Famous naturalists like Charles Darwin, Alfred Russel Wallace, and James Audubon shot, killed, and collected all manner of exotic creatures, in many cases showing a special appetite for brightly plumed birds and their colorful eggs. As recent books like *The Feather Thief* remind us, there are still those so obsessed with exotic bird plumage that they will break into museums to steal Wallace's trophies, which truly merit the label "rare birds."[12] At the time of his demise, the eccentric British millionaire Vivian Hewitt (Vivian Vaughan Davies Hewitt) owned half a million birds' eggs, including three of the seventy-five known eggs of the Great Auk, which had become extinct in the 1840s. In 2016, British ornithologist Tim Birkhead would publish a popular book on bird eggs entitled *The Most Perfect Thing*, in which he chronicled the exploits of obsessive collectors.[13] The perfect-egg theme is echoed in David Attenborough's documentary "Wonder of Eggs,"[14]

Konstantin Mikhailov was one of many of his generation whose love of biology and birds led him into the popular youthful pursuit of egg collecting. It was not until the latter part of the twentieth century that such collecting was frowned on, and in many places made illegal for most wild species and especially for protected species. Since many ornithologists had shown little or no interest in fossil eggs, this left a niche open for specialists like Karl and Konstantin, who found that these intriguing curiosities had been largely ignored, overlooked, and unstudied.

We already know that the truth of who laid what fossil egg can only be proven in cases where embryos are found and identified inside unhatched eggs, or occasionally as eggs inside a mother's body cavity. We also know that such finds are rare. For example, recall that poor *Oviraptor* was branded a thief and falsely accused of stealing what were incorrectly assumed to be *Protoceratops* eggs, when in fact it was guarding its own eggs. The proof of this correct reinterpretation was confirmed by *Oviraptor* embryos being found in the eggs.

Imagine, then, a library full of books that in 99 percent of cases contain only blank or smudged pages. However, the books are of different sizes and shapes and their covers are made of different materials and are differently ornamented and decorated. The covers reflect a certain craftsmanship and choice of materials, and one might wonder who made them. One might even organize them according to their bindings, as is done with multivolume editions.

In the classification of fossil eggs, the external shape is important. Some are large, some small, some rounded, some elongate, and so on. But even the external eggshell is not entirely external. We already know that some eggshell is thicker, some thinner; some are more multilayered and more or less heavily mineralized. It has internal structure, not just in differences between its layers and the composition and crystal "habits" of the minerals from which the layers are formed, but also in the geometry of the pores that perforate these layers. This is good news for the oologist wanting to find as many criteria as possible to help tell one egg from another.

It was a Russian paleontologist, Andrey Sochava, who in the late 1960s first took a close look at the structure of dinosaur eggshell that had been collected from Mongolia in the 1940s. She recognized and named "three distinct types of pore canals."[15] The long-winded terminology used to describe these different structures (for example, *multicanaliculate*—one of the easier terms to decipher, meaning "with multiple pore canals") is not a terminology with much formal biological meaning. This is because it does not really help us classify the whole egg.

It remained for a Chinese oologist, Zhao Zi-Kui, to take a broader, more holistic look at whole eggs: their shapes, or macrostructures; their microstructure (pore canals, ornament, and so on), best seen under the microscope; and their ultrastructure (for example, crystallographic properties), revealed only in the higher magnifications allowed by the electron microscope.[16] The terms proposed by Zhao (pronounced "jow" by Westerners) did not catch on immediately with non-Chinese paleontologists who did not speak or read Chinese, but they did follow the universally recognized and "formal" conventions that paleontologists use for naming fossil species, often in Greek or Latin. These conventions are explained in detail in almost every paleontology book, and although it is a challenge to do so in words of one syllable, the principle is simple. For example, as well as referring to man's

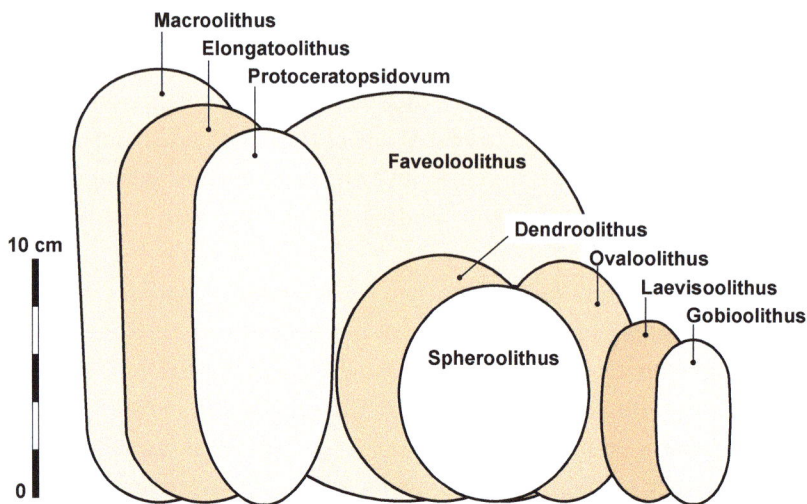

Eggs classified by shape according to the methods proposed by Chinese oologist Zhao Zi-Kui and colleagues.

best friend as "a dog," we can refer to dogs as members of the species *Canis familiaris* (always in formal italics), where *Canis* is the genus, a broader group containing wolves, jackals, and coyotes, and the family Canidae (no italics) is even broader, containing foxes (genus *Vulpes*). So, it is not surprising to learn that there are a number of species and genera (plural of *genus*) in the dinosaur family Tyrannosauridae, of which the genus *Tyrannosaurus* and the species *T. rex* are famous.

So, in 1979 Zhao named particularly distinctive and well-rounded spherical eggs to the oofamily as Spheroolithidae, containing the oogenus *Spheroolithus* which in turn contained three distinctive oospecies. Between 1975 and 1979, he also named oogenus *Elongatoolithus*, oogenus *Ovaloolithus*, and oogenus *Macroolithus*. No prizes for guessing the shapes of these various oogenera. Ooh ooh, I hear you say this egg-naming game is not so difficult. Zhao indeed pioneered a classification system that had merit for a certain simplicity. The year 1979 was a symbolic one for Karl, as this was when his first paper on fossil eggs appeared. He was then about to enter more deeply into a specialized field in which Chinese, Russians, Americans, Germans, and others would communicate in arcane technicalities, held together by the specialized conventions of classification in Greek and Latin.

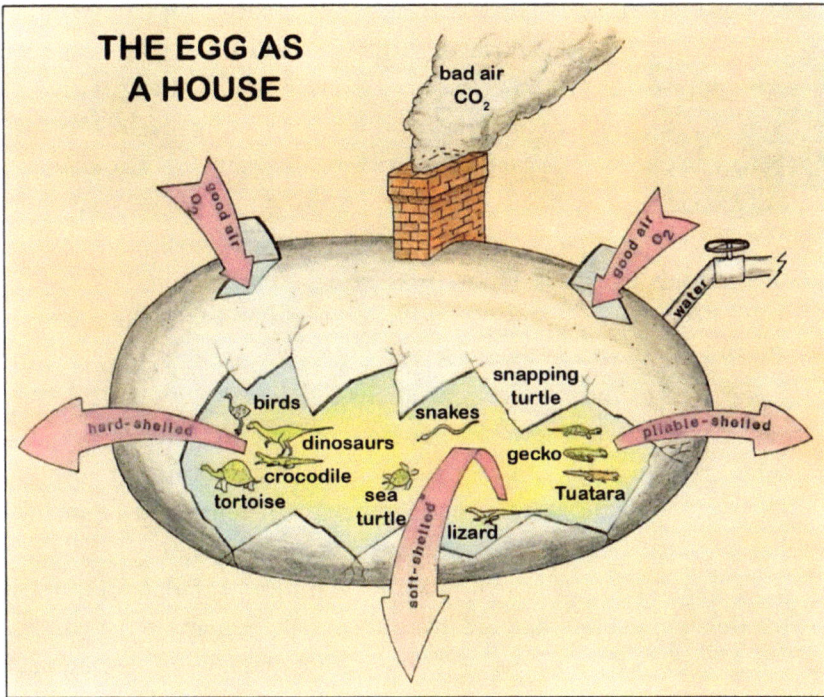

Karl used this simple "egg as a house" model to explain the function of eggshell to schoolkids. He made many such sketches, but this colored version, with caption, was published in the University of Colorado Boulder *Summit* magazine in 1990.[22]

A wise man, or perhaps a wise woman, once advised the scientific specialist to go out to the man or woman "in the street" and explain, as simply as possible, what their specialist research was about and why it was important. (And if you can't, ask yourself why?) Karl had the opportunity to do this explaining, though often to the child in the classroom rather than to those wandering the streets. His most famous analogy was the egg-as-a-house metaphor. "I explain to kids," he said, "that the egg is like a house where baby dinosaurs are protected from the outside world" (see chap. 10). The shell is like a screen to keep insects out, but the pores let fresh air with oxygen in the window and let the bad air (CO_2) out. Extending the metaphor back to Europe, he said that in almost any European country, you can say which country you are in by the way the houses are built; there are elongate, oval, round, and square houses. But that does not mean you know who the house belongs to, unless you happen to meet the owner. Likewise, you need to find the embryo in the egg to say who laid the egg. As far as extinct dinosaurs are concerned, their eggs have proved that most species were egg layers and did not give birth to live young, a fact that was not known with any certainly until the famous Mongolian discoveries less than a century ago. Dinosaur eggs also generally show that hatchlings were quite small and had to grow quickly in order to reach the large sizes typical of many species.

Why the study of eggs is important is perhaps a proverbial "no-brainer." Apart from the role that eggs play in the propagation of new life, they are common to all animal species and have their plant equivalent in spores and seeds. They are the link between generations and their evolution has allowed adaptation to different environments, from sea and fresh water to underground burrows and dry and arid lands. This is to say nothing of how eggs may be distributed externally in huge numbers, gathered together in smaller numbers at incubation sites such as nests, or guarded inwardly for internal gestation. Eggs are rich in protein and provide an almost endless food supply for countless omnivorous and carnivorous species. Humans in the Western world may eat mostly the eggs of a rather limited number of species, most notably the chicken, the basis of a huge food industry. However, humanity as a whole is more omnivorous, also partaking of the eggs of crabs and other invertebrates as well as the roe of fish such as the sturgeon, which is famously known as caviar.

Dealing specifically with eggshell and not the squidgy crab or fish roe variety, it is well known that eggshell can be an important indicator of the reproductive health of a species, and an indicator of ecological health. Infamously, the use of the pesticide DDT (dichloro-diphenyl-trichloroethane), an insidious colorless and odorless organochlorine, resulted in drastic thinning of eggshell in birds of prey at the top of the food chain. This put the iconic American bald eagle in danger of extinction, from which it recovered thanks only to a ban on most DDT use in the 1970s. The mechanism causing the thinning is complex but evidently is the result of reduced transport of calcium carbonate, the main eggshell ingredient, to the eggshell gland.

So, complex or not, the inconspicuous eggshell fragment may hold cryptic secrets that are only partially revealed, including clues to the ever-popular subject of dinosaur extinctions. When Karl passed away in 1996, he had a number of manuscripts in various stages of preparation, some of them

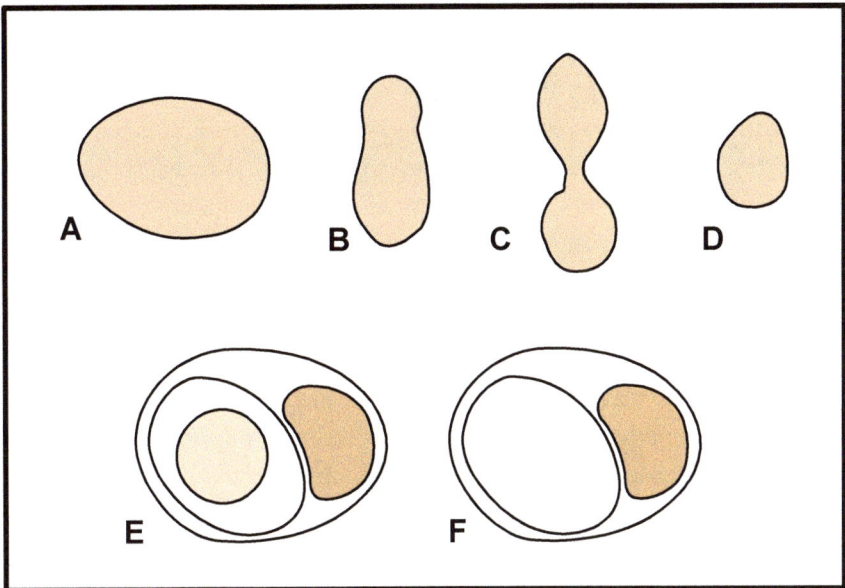

Pathological eggshells: external abnormalities. *A*, wrinkled surface; *B, C*, peanut and dumbbell shapes; and *D*, truncated shape. *E, F*, internal *ovum in ovo* abnormalities; *E* is a double yolk. Drawing modified after Hirsch (2001); see text and note 19.

involving his conscientious colleagues as coauthors. His legacy would not be extinguished with his death. Karl's first posthumous leap into the twenty-first century would come in 2001, with the publication of his forty-eighth paper, on "pathological" eggshell.[17] With Karl still a "star" of eggshell research, his name on the paper is marked with an asterisk and the somber notation "deceased." In this epistle from "the beyond," we are reminded that little is known in this field of eggshell pathology, although oddly shaped eggs have been of interest to science and to poultry farmers, if only as curiosities, since the curious first bothered to write about them. But, as is so often the case, explanations for such curiosities can be elusive, and such vague causes as "stress" are often cited with little conviction. In the case of chickens, oddly shaped eggs may be the result of "convulsive uterine contractions," which sound decidedly uncomfortable.[18] Presumably these might be abnormal contractions and not part of the normal egg-laying routine, since the results are wrinkly or misshapen, not the "perfect thing" over which Tim Birkhead waxed so lyrical.[19] But, as with so many things, pathologies may appear both external and/or internal. The egg may look perfect on the outside but be an "egg within an egg" (*ovum in ovo*, if you prefer the Latin). This happens in birds when the unshelled egg is sent back to the uterus, to meet the next one in line, before it descends again to the shell-secreting region.

We are again reminded of the bird–dinosaur connection that Karl mentioned regarding the Jurassic *Allosaurus* egg (chap. 12), which had been compared, for size, with the egg of the California condor. In this case, however, the comparison is based on eggshell structure, not size. Recall that the *Allosaurus* eggshell was extra-thick, having been held back in the oviduct shell-secreting region where it became multilayered. The opposite was the case with modern condors, which suffered a well-documented phase of eggshell thinning prior to the banning of DDT. Shells that were 0.7 mm thick pre-DDT thinned to less than half this size (0.3 mm) before DDT was banned in 1984. Even though these eggs were laid intact and could sometimes hatch, they could easily be crushed during incubation. This being said, there are rare examples of abnormal multilayered eggshells in birds.[20] The conclusion: the pathologies in modern and fossil eggs and eggshell are different and the causes are often unknown. It is possible that pathologies in late Cretaceous dinosaur eggs, just prior to their extinction, are a sign of some sort of environmental stress that caused eggshell thinning and hastened their demise. However, as Karl would pronounce from afar, one can only say that this "may have" been the case. Nice idea, but speculative, based on too little evidence, and "disputed" by many.[21]

None of this uncertainty means that fossil eggshell is not useful, for speculating about either the health, the pathologies, or the reproductive strategies of ancient egg layers. Due to its organic content, what a geologist might call "recent" or "sub-recent" eggshell has been used to date relatively young eggshell. Ostrich eggshells common at many sites in Africa and Asia have been used to date "recent" Ice Age deposits in the range of forty thousand to two hundred thousand years old.

As both Konstantin and Karl well understood, for every egg with an embryo or nest with a baby, there were thousands of unidentified eggs and tens of thousands of broken bits of eggshell—books with no pages and covers

with scrambled titles. There was still very little known about eggshell structure and what it meant in the ongoing quest to understand who laid what.
The quest is something like the detective work required to determine which
dinosaur or animal species or family was responsible for making a particular
type of track. What draws paleontologists like Karl and Konstantin to the
detective work that tries to make these difficult and often speculative connections is not easy to ascertain, even for the most ambitious psychiatrist or
philosopher of science. It is surely not the unrealistic hope that all answers
will be easy to find, or that a jigsaw known to have many missing pieces could
ever be complete. It is not the goal but the challenge that seems to be the mysterious ingredient driving human curiosity. Some people climb mountains
simply because they are there.

Because eggs and eggshells are not the actual dinosaurs—although one
might argue they are, in the rare cases when they contain an embryo—they
have been lumped with poop and tracks in different fossil categories. Thus,
eggs and eggshells, along with nests, as we have seen, were considered "trace
fossils." For this reason, Karl, Konstantin, and one of Konstantin's Russian
colleagues would participate in the First International Symposium on Dinosaur Tracks and Traces, in New Mexico in 1986. They, and two contributors
on the occurrence of dinosaur nests and egg clutches in India and China,
would be the "recognized" small delegation of egg experts, in a much larger
swarm of tracker delegates whose main interest was footprints. While Karl
would review the spotty distribution of eggshells across much of the globe,
the contribution of Konstantin and his comrade would focus on a dinosaur nest in Mongolia, which like some nests from Montana appeared to
show evidence of parental care. Perhaps more importantly, both Karl and
Konstantin were conveying a message that eggs, nests, and babies could be
found practically anywhere. Sometime in the late 1980s, the field of fossil egg
research (paleo-oology) had gone global. We have Americans, Mongolians,
Russians, Chinese, and Poles to thank for finding and describing so many
Central Asian nests and eggs. We can thank Zhao for his early classification
efforts and Karl and Konstantin for translating, refining, and disseminating
egg classification in the international scientific literature.

The globalization of fossil egg studies had, and still has, both upsides and
downsides. The good news is that increasing brain power is brought to bear
on a growing database and a difficult subject. The bad news is that efforts

to categorize, standardize, analyze, and make sense of the data inevitably generates much new, arcane, and often unpronounceable terminology— though as we have seen, Zhao's classification is a useful and common-sense classification of egg shape. The proverbial man or woman in the street knows the difference between a rugby ball (or American football) and a soccer ball!

Notes

1. T. I. Yusupova, "The Nonscientific Aspect of a Scientific Expedition: The Organization of the First Mongolian Paleontological Expedition, 1946, under the Leadership of I. A. Efremov," *Paleontological Journal*, 50, no. 12 (2016), 1298–1305.

2. K. Sabath, "Upper Cretaceous Amniotic Eggs from the Gobi Desert," *Paleontological Polonica*, 36 (1991), 151–192. See page 153. Also see See chapter 9, note 6, for a listing of some of Sochova's works.

3. I. A. Efremov, *The Road of Winds* [Russian edition]. Moscow, 1955. Also see I. A. Efremov, *Stories* (Moscow: Foreign Languages Publishing House, 1954). The book's 260 pages include eight stories, among them "The Shadow of the Past" (pages 11–54).

4. Efremov, "The Shadow of the Past," 10.

5. Efremov, "The Shadow of the Past," 16–17.

6. Efremov, "The Shadow of the Past," 27.

7. Efremov, "The Shadow of the Past," 240.

8. E. N. Kurochkin and R. Barsbold, "The Russian-Mongolian Expeditions and Research in Vertebrate Palaeontology, in *The Age of Dinosaurs in Russia and Mongolia*, ed. M. J. Benton et al. (Cambridge: Cambridge University Press, 2000), 235–255. See page 239.

9. See K. Sabath, "Upper Cretaceous Amniotic Eggs."

10. K. Sabath, "Upper Cretaceous Amniotic Eggs," 247.

11. K. E. Mikhailov, "Eggs and Eggshells of Dinosaurs and Birds from the Cretaceous of Mongolia," in *The Age of Dinosaurs in Russia and Mongolia*, ed. M. J. Benton et al. (Cambridge: Cambridge University Press, 2000), 235–255. See page 560.

12. K. W. Johnson, *The Feather Thief* (London: Penguin Books, 2019), 336.

13. T. Birkhead, *The Most Perfect Thing: Inside and Outside a Bird's Egg* (New York: Bloomsbury USA, 2016).

14. "Attenborough's Wonder of Eggs," *Natural World* (BBC Two, 2018–2019), https://www.bbc.co.uk/programmes/b09yj7dx.

15. K. Carpenter, *Eggs, Nests, and Baby Dinosaurs: A Look at Dinosaur Reproduction* (Bloomington: Indiana University Press, 1999), 148.

16. Z. Zhao and R. Li, "First Record of Late Cretaceous Hypsilophodontid Eggs from Bayan Manduhu, Inner Mongolia, *Vertebrata Palasiatica* 26 (1993): 107–115.

17. K. F. Hirsch, "Pathological Amniote Eggshell—Fossil and Modern," in *Mesozoic Vertebrate Life*, ed. D. H. Tanke and K. Carpenter (Bloomington: Indiana University Press, 2001), 378–392.

18. Hirsch, "Pathological Amniote Eggshell," 379.

19. T. Birkhead, *The Most Perfect Thing*.

20. F. D. Jackson and D. J. Varricchio, "Abnormal, Multilayered Eggshell in Birds: Implications for Dinosaur Reproductive Anatomy," *Journal of Vertebrate Paleontology* 23, no. 3 (2003): 699–702.

21. Hirsch, "Pathological Amniote Eggshell," 378.

22. J. Scott, "As Time Goes By," *Summit Magazine* (Boulder: University of Colorado, 1990), 18–20. Also see J. Scott, "Shell Game," *Summit Magazine* (Boulder: University of Colorado, 1990), 14–17.

CHAPTER SIXTEEN

A Man of Letters

A widower for twelve years since Hildegard had died, Karl passed away peacefully in bed in the summer of 1996 at the home of friends in the small town of Hurricane, Utah. The Hurricane location was perhaps symbolic of the storms of war that swept through his early life and the ill winds that had blown through his family life when his heart had troubled him, when Hildegard had contracted cancer, and when his adopted sons had disappointed a couple who had been unable to produce children of their own. Emily Bray would also recall Karl's determination as he collected eggshell in a Montana windstorm that threatened to asphyxiate them with suffocating blasts of stinging sand and dust.

Karl's egg studies had given him purpose and the satisfaction of contributing to a new field of paleontology. The first of his papers to appear after his death in 1996 was the first he coauthored with Konstantin Mikhailov and Karl's University of Colorado colleague Emily Bray. The first sentence of the paper contains this statement: "The tremendous increase in fossil egg and eggshell discoveries throughout the last decade necessitates the establishment of uniform methods for description and ... classification of fossil eggs."[1] The statement spoke to a dynamic and rapidly maturing scientific discipline and specialized community of scholars, and it could almost be edited into an epitaph for an egg scientist such as Karl.

Those who outlived Karl, including Emily, Konstantin, and many others (present authors included) would have time to reflect on the man they knew as a colleague and a friend. He offered food for thought for those interested in psychology and twentieth-century history. For those who knew him well, he was simply Karl, a "one off" character, perhaps a little larger than life, not least because of his strong accent and strong weather-beaten presence. His role as a pioneering paleontologist was perhaps secondary to the essential man. Thus, it would be mundane and uninspiring to merely list his awards and accolades.

One important facet of the essential Karl was his love of the outdoors, which dovetails seamlessly with all those activities that paleontologists call fieldwork and exploration. Since his teenage years he had been an explorer, and as a soldier he had spent months trudging under open, windswept skies and sleeping under the stars or tattered canvas. What paleontologists call "the field" was a regular escape destination. In the early days of exploration in Colorado he enjoyed the companionship of Hildegard and new friends in the rockhounding community. Despite providing help to colleagues from the university, he was perhaps, by definition, at least initially more amateur than professional, with a love for the wide-open spaces and the earth beneath his feet and dirt under his fingernails. But in Karl's case the amateur-professional distinction was largely academic: he could comfortably wear either label.

Inevitably, things changed after Hildegard died. He took her directives to his lonely heart and knuckled down to the business of professional paleontology, and sensibly he sought the company of friends and colleagues and, when necessary, communion with cognac. If loneliness was the ailment, companionship was the cure. These were ways to deal with the loneliness he so poignantly expressed a year after her death (chap. 9). It perhaps takes no great psychological insight to deduce that a man who had lost his mother as a child, had been the recipient of occasional traumatic abuse, had been wounded twice in war, and then had been subject to life-threatening imprisonment would suffer psychic scars. Add to this the trauma of a heart attack and the premature loss of Hildegard, his life's companion and link to his cultural roots, and one is forced to conclude that destiny had not dealt Karl the easiest hand. One could sometimes see the loneliness in his eyes, as if he was haunted and mildly puzzled by the hand that life had dealt him. It was a vulnerability that was hard to hide, but one that made him human, engendering empathy among friends, who offered him the balm of companionship. It was easy to attribute the scars to the slings and arrows of war and to the collateral-damage misfortunes he had endured, and so easy to have sympathy for a man who had had a tough life. But one also had respect for his integrity and tenacity. Sure, he covered his loneliness and angst with a certain bravado and impatience to get on with life, but he never fell into the trap of covering his insecurities with anger or boorishness. To the contrary, he mostly projected a friendly, even gentle conviviality. He dedicated his later years to his research with organized determination. He remained human, vulnerable, lovable, and imperfect.

By his final years Karl had become a globe-trotting researcher, attending international conferences. Konstantin remembers a conference in Frankfurt in 1992, also attended by Emily, where "some old man very politely even timidly said something to me. I did not understand at first . . . and only . . . maybe two minutes later understood that it is Karl Hirsch."[2] Thus, he met the seventy-one-year-old veteran of Operation Barbarossa. As was typical of the conference scene, they later

> drank . . . good German beer watching . . . the final match between Danish and German team . . . [in the] European Championship in 1992, and the Danish team beat the Germans and became the Champion of Europe, first time in . . . Soccer history. I was very happy for the Danes and remember that Karl was very complacent. . . . He was not a true fan of . . . soccer and, above all, took life as it is, without any reassessment of themselves, his nation, his social group, etc. He was absolutely free from all these biases.[3]

Karl may have been free from political and philosophical biases, or sensible enough not to talk about them among his culturally diverse friends, but he was often impatient, at least insofar as the time allotted to mortals is finite. Emily remembers him admonishing her for standing in line for beer when she and Konstantin accompanied him to the aforementioned celebrations in Germany. Much to the annoyance of some in front of him in the queue, Karl marched to the front of the line to demand instant service, leaving Emily to apologize on his behalf with the explanation, perhaps unexpected but semiplausible, "Don't worry, he's an old prison camp survivor." This surely reminds one of the expression "What does one have to do to get a beer around here?"

Like a long-distance runner who comes from behind, as he entered his seventies Karl appeared to sense that time was running out. He was impatient to delve into his research, and the field that he had a significant hand in shaping. But he sought out the likes of Emily as colleagues and companions. It was a "need" that helped fill a void of loneliness. She and Karl would work in the field together, sometimes in foul windstorm weather that would send everyone but Karl and her scurrying for shelter, with only Emily staying to "grit it out"—quite literally, to match his drive and endurance. He was on a late-life mission. As Emily would recall, Karl consciously sought female company (chap. 9), perhaps to avoid confrontation with male competitors as had sometimes happened in the military sphere. Perhaps in his subconscious he preferred to remember those nurse angels, rather than enemy soldiers or corrupt officers. Perhaps he sought a species of father-daughter relationship he had been denied when Hildegard had decided they should adopt boys, not girls. His deeper needs were perhaps not easily explained, or perhaps not necessary to explain, at this stage in his life. He surely wanted to make the most of his allotted time and die with his boots on. His fear of aging, some friends would say, was the fear of losing his mental faculties.

When Konstantin came to Colorado in 1994, he and Emily and Karl would head for the proverbial field with Karl's dog Maggie, making an international quartet. Konstantin wrote,

> At that time Karl was very lonely in his personal life (it was impossible not to see it) and I felt that he is even happy that someone (his "younger friend") can share with him all the breakfasts and dinners, that he can recall with me his being in Russia, in a very peculiar state. I felt that I can pay him back for his hospitality only listening to him, supporting memories of his being young man with his particular connection with Russia, in such a special way. He was really happy (I remember!) when before my departure we did not go to cafe but arranged simple Russian-way supper with vodka, boiled potatoes and salty cucumbers. I could see some quiet joy in his eyes. That evening he told me about his life in prison, how he hated the war; he also tried to sung (*sic*) Katyusha song and wanted to hear again Garmoshka (Russian version of accordion which was broadly in use in the time of the War).[4]

This was Konstantin's second, poignant, and warmly empathetic impression. Karl repaid the companionship with equal warmth, insisting on paying for most meals, as Konstantin was, after all, the "invited" guest. There was more to this kindness than at first met the eye. Karl had in fact helped behind the scenes to procure a visa for Konstantin, who was at the time enjoying a research fellowship in England, where he and Karl had met on another occasion to drink vodka and eat potatoes with British students. The chronology seems to have been beer-snatching in Germany, vodka and potatoes in England, and then more vodka and potatoes in Colorado.

Picture of Karl with Emily in the Colorado high country. His dog Maggie rests at his feet. Photo courtesy of Konstantin Mikhailov.

Karl had also found funds to support a Wild West excursion that took them on a four-week tour to western Colorado, Egg Mountain, Yellowstone, Bryce Canyon, and Karl's mountain cabin. One ventures to suggest that Karl had learned some savvy requisition tricks during the war that served him well when it came to creatively supporting his research efforts. Again

Konstantin remembers, "[In the] evening Karl often invited me to Mexican restaurants or to small Italian I (. . . he did not like dining alone at home). It was fantastic time of my "summer vacation" in the Western Interior. And first time of acquaintances with its nature and American people whom we met during our trips. Thanks to Karl. It was his gift. And one of my happiest moments in life."[5]

One of the fruits of the camaraderie enjoyed by the Karl-Konstantin-Emily excursion trio was the aforementioned joint paper with a relatively simple title: "Parataxonomy of Fossil Egg Remains (Veterovata): Principles and Applications," published in 1996, the year Karl died.[6] The title, short by most specialist standards, could be paraphrased as "basic principles for naming the eggs of fossil vertebrates." It can be seen as a fitting tribute to his mature role in establishing a coherent international vocabulary in the specialized field to which he had happily dedicated his energies.

In a biography dedicated to the memory of Karl, both as man and as scientist, it is not always easy to separate these facets of a life. Biographers of scientists like Darwin, Wallace, Einstein, or other lesser scientific luminaries often deal with individuals steeped in scientific pursuits from early adulthood. In Karl's case, it is easier to separate the chapters of his Old World life in Germany and Soviet prison camp and his New World life in America as starkly compartmentalized by geography and time. However, there are other dimensions, and Karl inevitably carried his fundamental character and his physical and psychological wounds from Old World to New, even as the latter healed at least a little. It is easy to applaud, reward, and honor those who have overcome hardship and life-threatening trauma and have made conscious efforts to move on positively, to improve their lives and those of friends and family. Karl was in his own mind no medal-winning hero. It is always for others to bestow awards, and they did. Karl was philosophical enough to recognize that the slings and arrows of outrageous fortune visit each one of us unpredictably, and that existence has both its sad and tragic sides and its joyful and humorous ones. He had survived, and this was cause for optimism and getting on with life, as he would advocate to friends. This optimism was at least partly born of his experience with Russian prison guards. Telling how they only half-heartedly knocked prisoners around—pretending to punish the "enemy," sometimes with a wink that said "We guards are also Siberian exiles"—would bring a smile to Karl's face. He recognized the empathetic gesture and appreciated how they allowed prisoners smoke breaks, which ironically gave smokers respites that nonsmokers could not enjoy. If asked to divulge his autobiographical memories, it was these glimpses of humane behavior, rather than war trauma, that he chose to recount. Years later, returning empty-handed from a hunting trip, where his shots had failed to find their target, he would joke that "It is no wonder we lost the war." Shoveling heavy snow from his driveway in Denver, he would chuckle and say, "Just like in Russia, except that I can stop and go in for a cup of coffee whenever I want."

From a historical and political perspective, it is rather obvious that a German in postwar America would not likely receive honors as a combat veteran. (He had already received the Iron Cross equivalent from the Third Reich.) The honors bestowed upon him were purely academic: the Paleontological Society's prestigious Strimple Award, for amateurs who have furthered the field of paleontology, and the honorary degree of Doctor of Humane Letters from the University of Colorado.[7] Both awards were bestowed in 1990. Karl made his journey alongside and with the help of friends and colleagues, many of them present to see the awards given. Notwithstanding his scientific dedication and drive, it is never for the individual to measure his or her own contributions. It is for biographers, historians, and scientific associations to assess an individual's impact in academia and society and bestow the appropriate recognition.

As biographers we would be remiss if we did not mention Karl's friend and colleague Dr. Hans-Peter Schultze, an eminent paleontologist, who had been born in 1937 in a part of Germany (now in Poland) not far from where Karl had been raised. Like Karl, Hans-Peter had emigrated, in 1978, to the western USA, taking positions as Curator of Paleontology at the Museum of Natural History and associate professor at the University of Kansas. It was in this decade that Karl had begun to take scientific paleontology seriously and publish his first papers. Karl would drive east to Kansas, the land of Oz, to visit Hans-Peter as often as he would drive similar half-day distances to field sites in states north and west of Colorado. In Kansas, he and Hans-Peter would reminisce about the old country and its history, and discuss their various interests in life and professional paleontology. Hans-Peter would write Karl's obituary for the Society of Vertebrate Paleontology, the field's premier institution. He would venture beyond mere scientific facts to describe Karl as one who

> tried to pass on his knowledge to younger people and scientists . . . [helping] young students and scientists like Emily Bray and Darla Zelenitsky to grow from his knowledge. Many colleagues from overseas visited him . . . [as] . . . a friendly host. . . . He also impressed everyone as a friendly, helpful, and thoughtful person. He helped many people through difficult times in their lives. He had the capacity to listen to others, to let others express their problems and pains. He had based his principles on a lifelong experience [and] . . . tried to help others to understand their own situations before they made any decisions. The thoughtful procedure was the principal guidance in his research. He preferred and tried to convince his "pupils" to do a careful study before publishing anything.[8]

One cannot ask for a better encapsulation of how Karl blended his scientific dedication with the humane and empathetic side of communication with friends and colleagues. Having said this, we biographers were privileged to have been given the equally heartfelt comments of his friend and student Darla Zelenitsky, now a leader in the field of eggshell research. She writes of his legacy and the many papers published since Karl's passing, and how many have cited his work and have even named new eggshell types in his honor. These people, she says, are Karl's "descendants." Surely an apt evolutionary metaphor. She writes that Karl's "connection with people, ability to bring people together, and willingness to help others in research made the world of egg research, where people were widely distributed, much smaller at a time before internet." Whether on the phone with Darla, on the road to Kansas, Nebraska, Utah, or Montana, on a visit to Germany or England with colleagues like Emily and Konstantin, Karl played a vital networking role beneficial to the intellectual connective tissue that helped make the world of egg research coherent. Darla recalls that Karl "introduced me to everyone in his egg world . . . making collaborations . . . so much easier and productive . . . he felt collaboration and openness were important for the field. He was crucial and instrumental to the future of research on eggs. It was *him*, not his friends and collaborators, that did this for the future of the field . . . not anyone [else] . . . *it was Karl!*"[9] This ringing endorsement underscores the more nuanced and fundamental aspects of what makes science what it is supposed to be: universal, open, communicative, and synergistic down through the generations.

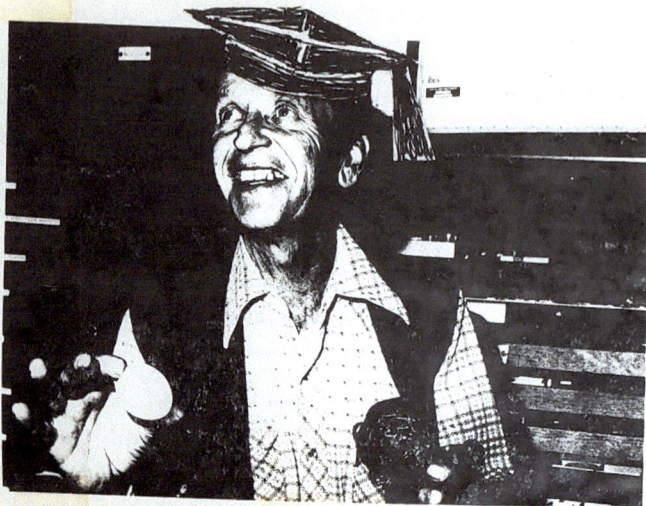

Lieber Karl!
Zur Verleihung des
Doktorhutes
gratulieren wir Dir
sehr herzlich!
Wir sind stolz auf Dich!

Anita und Hans-Jürgen
und auch
Anni, Clemens, Martina und Manuel

Berlin, den 11. 8. 1990

Facing, Consistent with the "Man of Letters" theme begun in chapter 12, Karl's family was proud of his achievements and sent him this homemade greeting. In translation, the message reads, "We congratulate you very much on being awarded the doctoral hat. We are proud of you. Anita and Hans-Jürge. Anni, Clemens, Martina and Manuel. Berlin, 11.8.1990." The hat is sketched onto the photo! University of Colorado Museum archives.

It is a tribute to Karl and to the esteem of his colleagues that the last ten of his scientific papers were published posthumously, and it is perhaps an odd fact that the very last paper dealt with eggs from a site in the former Soviet territory of Kazakhstan, geographically closer than any other Karl had studied to where he had been held prisoner in the heart of Asia. Coauthors, some he had mentored, rallied to ensure that his ongoing work was finished. We recall that as early as 1986, ten years before his death, he was representing his field alongside Konstantin on the international stage amid a congress of dinosaur trackers, although he and Konstantin did not meet at this time. Five years after the tracker symposium proceedings had been published by Cambridge University Press, in 1989, Cambridge's venerable publishers put their academic stamp on a 1994 book entitled *Dinosaur Eggs and Babies,* which Karl edited with colleagues Ken Carpenter and Jack Horner.[10] The book was dedicated to Bob Makela, Jack's beloved field crew chief, who had died in 1987 in a road accident in Montana.[11] The book, to which Karl co-contributed both the Introduction and the Prospectus, as well as his own paper on Jurassic eggshell, is symbolic of global collaboration and the arrival of the field in the rarified academic atmosphere of a four-hundred-year-old publishing house in Cambridge, England, far from the arid, eggshell-strewn badlands of Montana or Mongolia.

Early in his life, Kenneth Carpenter—everyone calls him Ken, except Karl who called him Kenny—had aspired to be a paleontologist working for what is now the Denver Museum of Nature and Science, Colorado's largest natural history museum. Through hard work and dedication Ken would achieve this goal. As a geology undergraduate at the University of Colorado he began collecting the fossils of dinosaurs and marine reptiles, even though the University of Colorado Museum tried to dissuade him from excavating too many, using the age-old excuse of limited space and resources. Uncertain of the likelihood of a high-level curatorial, research position in paleontology or in Colorado, he took on a series of hands-on natural history museum jobs involving exhibits and fossil preparation. Intellectually curious and with wide-ranging interests in field excavation, laboratory preparation, restoration, artistic reconstruction, and exhibits, he would eventually get his coveted Denver Museum job and complete a PhD at the University of Colorado, by which time he had racked up several hundred scientific publications.

Ken and Karl had launched on parallel paleontological paths at the University of Colorado at about the same time. Ken enrolled formally as a student, while Karl studied informally in the adult-education program under the tutelage and sympathetic guidance of life-long mentor and friend Judith Harris, from whom he took scientific advice and a class in vertebrate paleontology. In time, their paleontological paths would converge in the study of eggs and eggshells, thanks largely to Ken's wide-ranging interests in dinosaurs and the community's growing interest in dinosaurian biology and behavior.

While still a student collecting fossils for a reluctant University of Colorado Museum, in the early 1980s, the infamous Professor Robert Bakker, later known as Dr. Bob, a self-appointed dinosaur heretic and "enfant terrible," albeit with impressive paleontological credentials and a deep knowledge of dinosaur anatomy, packed up his professorship at Johns Hopkins University and headed west to hang his famously tattered straw hat at the University of Colorado Museum. Dr. Bob could claim to be the spokesman, labeler, and cheerleader for, if not the architect of, the "Dinosaur Renaissance" in the conceptually revolutionary late 1960s.[12] He could not claim to be the Joseph Wortmann Chair of Paleontology, an entirely fictitious position he made up to perpetuate his enfant terrible image and to irritate the university powers that be. Thus, the University of Colorado Boulder campus, with Ken, Karl, Dr. Bob, Judith, and others, though not necessarily in coordinated unison, would become an enclave for one of the West's more heterodox centers of dinosaur research. Few aspects of dinosaur paleobiology—whether eggs, nest, babies, reproduction, growth, locomotion, track-making (a specialization of the University of Colorado Denver campus), biomechanics, social behavior, or extinction—were not of keen interest or potential subjects for serious research, and a smattering of wild speculation. Dinosaur Renaissance was in the air, and Karl was destined to be part of the movement.

Jump ahead to the mid-1990s when *Dinosaur Eggs and Babies* was published, with the introductory statement penned jointly by Ken, Karl, and Jack (Horner) that "we are currently enjoying a new Golden Age of Dinosaur Paleontology."[13] Their introduction ended with reference to the "important collections made by the Central Asian Expedition" and the pithy remark that "We feel it appropriate to close the Introduction by briefly . . . [noting that] . . . despite the importance of these eggs, they have been treated as curiosities"[14] and only briefly described. *Dinosaur Eggs and Babies* was a landmark of sorts and a credit to Ken's initiative as coeditor, honed by his experience editing other dinosaur volumes and observing the impact of the work done by Karl and his colleagues in the field at large. Yes, it was a landmark, but it was by no means the last word on the subject of these oological curiosities. A review of the book in the scientific journal *Nature* was cleverly entitled "Dinosaur Growth Industry," and it stated that the book "succeeds admirably in summarizing the current state of knowledge while drawing attention to how much remains unknown."[15]

All business, Ken elaborated his *Dinosaur Eggs and Babies* knowledge into his sole-authored *Eggs, Nests, and Baby Dinosaurs: A Look at Dinosaur Reproduction*, "dedicated to my dear friends" Karl F. Hirsh, "who taught me about fossil eggs," and John R. Horner, "who taught me about baby dinosaurs."[16] As the reader will note, in the dedication of this book the favor is returned to Ken and colleagues who helped Karl, as he helped them: more global cooperation. The 1990s will be remembered as a landmark decade in which Karl, Konstantin, Emily, Ken, and colleagues conducted, published on, and synthesized much original, new fossil egg research, bringing the specialized work of previous decades out of the shadows. Besides being the decade that began with Karl receiving the Paleontological Society's prestigious Strimple Award, in 1990, it was also the decade in which Karl helped stimulate the next generation's interest and motivate rising stars in the world of egg research.

If Karl had survived into his eighties or nineties he would have seen papers published that established new connections between eggs and egg layers, as in the examples of eggs discovered inside the fossilized body cavities of their parents. He would have learned of large sauropod nest colonies in South America as impressive as those of Montana's duck-billed dinosaurs.[17] He might also have lived to see recent reports that some dinosaurs had soft-shelled eggs, and witnessed an explosion of literature on the reproductive biology of extant and fossil avian and nonavian dinosaurs and various reptile groups.[18] Today there is even evidence that nonavian theropod dinosaurs engaged in courtship behavior that ornithologists refer to as lekking, "nest scrape display" or "pseudo nest building" sites known as "display arenas "prior to the egg-laying and incubation stages of the breeding cycle.[19] Such new contributions often build on Karl's seminal pioneer work in the field of egg research.

Karl passed away mid-decade, on June 1, 1996, with research ongoing as Ken worked on *Eggs, Nests and Baby Dinosaurs*. The book would underscore the decade's significance in the annals of fossil egg research. The dedication to Karl, and the 1999 publication date, in the final year of the millennium, would bring closure to the century in which Karl had lived and made his mark. The egg research foundation that he laid with colleagues and disciples before the dawn of the present millennium would help further the incubation and fledging of this extraordinarily intriguing field of evolutionary paleobiology. We find that the answer to the perennial conundrum "Which came first?" was, in the case of Karl's endeavors, the egg!

Notes

1. K. E. Mikhailov, E. S. Bray, and K. F. Hirsch, "Parataxonomy of Fossil Egg Remains (Veterovata): Basic Principles and Applications. *Journal of Vertebrate Paleontology* 16, no. 4 (1996): 763–769.

2. Reminiscences of Konstantin Mikhailov sent to Martin Lockley (December 2018) via email.

3. Reminiscences of Konstantin Mikhailov.

4. Reminiscences of Konstantin Mikhailov.

5. Reminiscences of Konstantin Mikhailov.

6. Mikhailov et al., "Parataxonomy of Fossil Egg Remains."

7. For more information about the Strimple Award, see appendix C. Also, for more information about Karl's Honorary University of Colorado Boulder PhD, see appendix B.

8. H-P Schultze, "Karl F. Hirsch 1921–1996 (Obituary)," *Society of Vertebrate Paleontology News Bulletin* 168: 70–72.

9. Written communication from Darla Zelenitsky to the authors of this biography.

10. D. Quammen, "Friendships in Stone," *Outside Magazine* (November 1987): 23–26. This article is a memorial to Bob Makela, friend to both Karl Hirsch, with whom he shared a birthday, and Jack Horner. Makela died in a car accident in 1987.

11. K. Carpenter, K. F. Hirsch, and J. R. Horner, "Introduction," in *Dinosaur Eggs and Babies*, ed. K. Carpenter, K. F. Hirsch, and J. R. Horner (Cambridge: Cambridge University Press, 1994).

12. R. T. Bakker, "The Dinosaur Renaissance," *Scientific American* 232 (1975): 58–72.

13. Darla Zelenitsky, written communication to the authors of this biography.

14. Darla Zelenitsky, written communication.

15. A. Milner, "Dinosaur Growth Industry," review of *Dinosaur Eggs and Babies*, ed. K. Carpenter, K. F. Hirsch, and J. R. Horner, *Nature* 370, 1994.

16. Darla Zelenitsky, written communication, 18.

17. L. Chiappe and L. Dingus, *Walking on Eggs: The Astonishing Discovery of Thousands of Dinosaur Eggs in the Badlands of Patagonia* (New York: Scribner, 2001).

18. M. Norell et al., "The First Dinosaur Egg Was Soft," *Nature* 583 (2020): 406–410.

19. The first report of courtship display traces made by Cretaceous theropod dinosaurs is the first physical evidence that some dinosaurs behaved like modern ground-nesting birds. Such traces indicate that nesting sites must have existed nearby, even if they have not been found or preserved. M. G. Lockley et al., "Theropod Courtship: Large Scale Physical Evidence of Display Arenas and Avian-Like Scrape Ceremony Behaviour by Cretaceous Dinosaurs," *Scientific Reports*, 6, no. 18952 (2016).

Egg Money

Is Karl's collection of eggshell, microscope slides, photographs, tiny electron microscope specimens, and data sheets worth almost a quarter million dollars? Conscientious paleontologists would surely say yes, and surely worth much more! More likely they would say, "Silly question, it impossible to put a price on such things." Whatever one's opinion, paleontological collections are a nonrenewable resource. Here are a finite number of eggs of extinct dinosaurs: no new ones are being laid. Find the only fossil of a rare dinosaur, then lose or destroy it and it would have become extinct twice!

To answer the appraisal question, the Biological Research Collections Program of the National Science Foundation thought the collection was worth $237,945. This was not the intrinsic, retail, wholesale, or black-market value, but the perpetuity and annals of paleontology value, of a properly curated collection: not the collection itself. It is what the government-funded agency was willing to pay the University of Colorado Museum of Natural History to catalog and preserve, that is, curate, the collection in what is called an "accredited," that is, government-approved, repository.

Obvious as it may be to some readers, when valuable specimens are given to a museum the donation, or acquisition, has to be recorded and the specimens have to be appropriately labeled and stored with all relevant or ancillary data. This process of "curation" involves a curator who makes the executive decisions—for example, what to accept in the first place, and whether some of the specimens should be exhibited or perhaps loaned to other scientists for further study. These days, descriptions, images, and interpretations may also be posted online. The collection manager has an important role, rather like bookkeeping or accounting.

Imagine for a moment that you are a curator who has decided to accept the now-enshrined Karl Hirsch Fossil Eggshell Collection for the University of Colorado Museum of Natural History. You made this decision because Karl had worked at the museum as a research associate, and because Karl had been a Colorado resident it is a "no-brainer" for you to say that this is where the collection belongs, especially when curators describe it on the museum website as "exceptional . . . one of the most informative collections of eggshells in the world, with over 3,000 catalogued fossil and recent specimens from over thirty countries, and ranging in age from the Jurassic to the recent." Even assuming that someone will fund the work that is entailed, and happily the National Science Foundation did, what is involved in taking in the collection?

It turns out that there is much more to the collection than three thousand relatively small eggshell fragments. Quoting from the museum's final report, there is "auxiliary data in the form of approximately 35,000 photographic negatives, 3,000 thin sections, and 20,000 data pages." The report goes on to detail the many people who participated in the work, "including a curator, a collection manager, two research associates, two graduate and thirteen undergraduate students, a local high school teacher, and colleagues from the University of California Museum of Paleontology, Montana State University, and a software development firm (Whirl-i-gig)."

The National Science Foundation seems to have got quite good value for its grant money. Crude arithmetic suggests that if someone were to simply spend half the money on the six thousand eggshell and microscope specimens, and the other half on the photographs, the specimens would cost less than $20.00 each and the photographs less than $4.00 each. Of course, this is not how it works, and we must factor in at least twenty people who worked on the project, and unknown doses of volunteer effort.

University of Colorado curator Dr. Karen Chin was responsible for obtaining the $237,945 grant and coordinating these curation efforts between 2008 and 2011. She had known Karl personally, as had Emily Bray and others who worked on the project, and they all helped bring the collection into the University of Colorado fold. The National Science Foundation guidelines that hover over collections management projects steer museums to do more than put valuable specimens and archives safely under lock and key. They also want museums to make the specimens, but also, more generally, the "significance" of the specimens, available to a wide audience. There are always politicians who will get upset if they see a government agency funding some obscure project that seems to have no obvious benefit to society. So why are fossil eggshells important?

Well, first, the same government that issues modest amounts of research and collections management grant money has also declared it illegal to collect such specimens without putting them in a recognized museum. (So, no further justification needed!) The same politicians, who might question spending money on fossil eggshell research, would be equally indignant if the specimens were thrown away after money had been invested in their care and study. They might also be indignant if, having heard of the work of a scientist like Karl Hirsch, connected to the University of Colorado, they found that it was impossible to find the specimens, exhibits, or evidence of his groundbreaking work. In short, the justification for museum collections and scientific endeavor boils down to four words: "preserving the national heritage." Second, though other justification is hardly necessary, there is the educational value of all fossils for teaching students about prehistory and getting them interested in science in general. Karl had done his share of outreach in this department, taking his egg-as-a-house story out to schools in the area. Dinosaurs and their eggs and babies come up trumps in this arena, with no shortage of student interest in birds, crocodiles, turtles, lizards, and snakes. Cute babies are simply the icing on the cake.

Karl died more than a decade before Karen obtained funding to put his collection in order. Although born in the typewriter-and-radio generation, he had, by the early 1990s, become a member of the electron microscope-and-email generation. However, he did not live to see his collection finally curated, or to see the "Karl Hirsch and the Hirsch Eggshell Collection" web page. Web browsing was in its infancy when Karl died, and Google did not exist as a much-used word in the global vocabulary. Two decades later, copious information on paleontology collections is available online. The University of Colorado Museum collections are part of the Paleontology Portal system, linking more than a dozen museum collections. Karl and his legacy are with us still in the twenty-first century age of Whirligig and Google. The information is just a click away: https://ucmp.berkeley.edu/science/egg shell/eggshell_hirsch.php

The enshrining of the Karl Hirsch Eggshell Collection at the University of Colorado, the state in which Karl spent almost half his years, seems a fitting end to Karl's whirligig life. Somewhere among the papers and photographs there may be a smidgeon of Hildegard's DNA, testimony to the supporting role she played in making Karl "the dinosaur egg man," and surely there are smudges of the DNA of his friends and colleagues all over the collection, both literally and figuratively. There may also be a sheet somewhere with a subtle, golden, arc-shaped trace that a forensic detective might identify as a strong black coffee, beer, or cognac stain, perfectly crescentic in form like the eggshell slices to which Karl had dedicated himself.

APPENDIX A

Bibliography of Karl F. Hirsch

Bray, E. S., and K. F. Hirsch. "Eggshells from the Morrison Formation." In *The Upper Jurassic Morrison Formation: An Interdisciplinary Study,* edited by K. Carpenter, D. Chure, and J. I. Kirkland. *Modern Geology* 23 (1998): 219–240.

Carpenter, K., K. F. Hirsch, and J. R. Horner. "Introduction." In *Dinosaur Eggs and Babies,* edited by K. Carpenter, K. F. Hirsch, and J. R. Horner. Cambridge: Cambridge University Press, 1994.

Carpenter, K., K. F. Hirsch, and J. R. Horner. "Summary and Prospectus." In *Dinosaur Eggs and Babies,* edited by K. Carpenter, K. F. Hirsch, and J. R. Horner, 366–370. Cambridge: Cambridge University Press, 1994.

Dantas, P. M., J. J. Moratalla, K. F. Hirsch, and V. F. Santos. "Mesozoic Reptile Eggs from Portugal. New data." *Dinosaurs and Other Fossil Reptiles of Europe.* Second Georges Cuvier Symposium, September 8–11, 1992, Montbéliard.

Hayward, J. L., K. F. Hirsch, and T. C. Robertson. "Rapid Dissolution of Avian Eggshells Buried by Mount St. Helens Ash." *Palaios* 6 (1991): 174–178.

Hirsch, K. F. "Contemporary and Fossil Chelonian Eggshells." *Copeia* (1983): 382–397.

Hirsch, K. F. "Die Ammoniten des Pierre Meeres (Oberkreide) in den Westlichen USA." *Der Aufschluß* 26 (1975): 102–113. Heidelberg.

Hirsch, K. F. "Fossil Crocodilian Eggs from the Eocene of Colorado." *Journal of Paleontology* 59 (1985): 531–554.

Hirsch, K. F. "The Fossil Record of Vertebrate Eggs." In *The Paleobiology of Trace Fossils,* edited by S. K. Donovan, 269–294. Chichester, UK: Wiley, 1994.

Hirsch, K. F. "Fossile Eier—ja Oder Nein?" *Aufschluss* 38 (1987: 253–258. Heidelberg.

Hirsch, K. F. "Interpretations of Cretaceous and PreCretaceous Eggs and Shell Fragments." In *Dinosaur Tracks and Traces,* edited by D. D. Gillette and M. G. Lockley, 89–97. Cambridge: Cambridge University Press, 1989.

Hirsch, K. F. "A Look at Pathological Amniote Eggshell—Fossil and Modern." *Journal of Vertebrate Paleontology* 9, suppl. 3 (1989): 81.

Hirsch, K. F. "Not Every 'Egg' Is an Egg." *Journal of Vertebrate Paleontology* 6 (1986): 200–201.

Hirsch, K. F. "The Oldest Vertebrate Egg?" *Journal of Paleontology* 53, no. 5 (1979): 1068–1084.

Hirsch, K. F. "Parataxonomic Classification of Fossil Chelonian and Gecko Eggs." *Journal of Vertebrate Paleontology*, 16, no. 4 (1996): 752–762.

Hirsch, K. F. "Pathological Amniote Eggshell—Fossil and Modern." In *Mesozoic Vertebrate Life*, edited by D. H. Tanke and K. Carpenter, 378–392. Bloomington: Indiana University Press, 2001.

Hirsch, K. F. "Upper Jurassic Eggshells from the Western Interior of North America." In *Dinosaur Eggs and Babies*, edited by K. Carpenter, K. F. Hirsch, and J. R. Horner, 137–150. Cambridge: Cambridge University Press, 1994.

Hirsch, K. F., and J. Bowles. "Early Eocene Cranelike Eggs?" *Proceedings 1978 Crane Workshop*: 211–216. Rockport, Texas.

Hirsch, K. F., and E. Bray. "Spheroidal Eggs (Avian and Chelonian) from the Miocene and Oligocene of the Western Interior." *Journal of Vertebrate Paleontology* 7, suppl. 3 (1987): 18A.

Hirsch, K. F., and E. S. Bray. "Spheroidal Eggs—Avian and Chelonian—from the Miocene and Oligocene of the Western Interior." *Hunteria* 1, no. 4 (1988): 1–8.

Hirsch, K. F., and J. Harris. "Fossil Eggs from the Lower Miocene Legetet Formation of Koru, Kenya: Snail or Lizard?" *Historical Biology* 3 (1989): 61–78. Chur, Switzerland.

Hirsch, K. F., A. J. Kihm, and D. K. Zelenitsky. "New Eggshell of Ratite Morphotype with Predation Marks from the Eocene of Colorado." *Journal of Vertebrate Paleontology* 17.

Hirsch, K. F., and R. Kohring. "Crocodilian Eggs from the Middle Eocene Bridger Formation, Wyoming." *Journal of Vertebrate Paleontology* 12, no. 1 (1992): 59–65.

Hirsch, K. F., L. Krishtalka, and R. Stucky. "Revision of the Wind River Faunas, Early Eocene of Central Wyoming, Part 8: First Fossil Lizard Egg (?Gekkonidae) and List of Associated Lizards." *Annals of Carnegie Museum* 56, no. 12 (1987): 223–230.

Hirsch, K. F., and L. F. LopezJurado. "Pliocene Chelonian Fossil Eggs from Gran Canaria, Canary Islands." *Journal of Vertebrate Paleontology* 7, no. 1 (1987): 96–99.

Hirsch, K. F., and M. J. Packard. "Review of Fossil Eggs and Their Shell Structure." *Scanning Microscopy* 1 (1987): 383–400.

Hirsch. K. F., and Quinn, B. "Eggs and Eggshell Fragments from the Upper Cretaceous Two Medicine Formation in Montana." *Journal of Vertebrate Paleontology* 8, suppl. 3 (1988): 17A.

Hirsch, K. F., and B. Quinn. "Eggs and Eggshell Fragments from the Upper Cretaceous Two Medicine Formation of Montana." *Journal of Vertebrate Paleontology* 10, no. 4 (1990): 491–511.

Hirsch, K. F., and H. L. Robinette. "Looking into a Madstone." *Colorado Outdoors* 35, no. 2 (1986): 23.

Hirsch, K. F., K. L. Stadtman, W. E. Miller, and J. Madsen. "A Pathological Jurassic Dinosaur Egg Containing an Early Stage Embryo from Central Utah." *Journal of Vertebrate Paleontology* 8, suppl. 3 (1988): 17A.

Hirsch, K. F., K. L. Stadtman, W. E. Miller, and J. H. Madsen. "Upper Jurassic Dinosaur Egg from Utah." *Science* 243, 4899 (1989): 1711–1713.

Hirsch, K. F., D. Thies, J. Vespermann, and W. Weinrich. "Blick ins Innere von Dinosauriereiern." *Spektrum der Wissenschaft* 10 (1994): 24–32. Weinheim.

Hirsch, K. F., R. Young, and H. J. Armstrong. "Eggshell Fragments from the Jurassic Morrison Formation of Colorado." In *Paleontology and Geology of the Dinosaur Triangle, Guidebook*, edited by W. Averett, 79–84. Grand Junction: Museum of Western Colorado, 1987.

Hirsch, K. F., and D. K. Zelenitsky. "Dinosaur Eggs: The Identification and Classification." In *Dinofest International Symposium, Programs and Abstracts*, edited by D. L. Wolberg, and E. Stump, 61. Tucson: Arizona State University, 1996.

Kohring, R., and K. F. Hirsch. "Crocodilian and Avian Eggs and Eggshells from the Eocene of the Geiseltal, Eastern Germany." *Journal of Vertebrate Paleontology* 16, no. 1 (1996): 67–80.

Lucas, S. G., E. S. Bray, R. J. Emry, and K. F. Hirsch. "Dinosaur Eggshell and the Cretaceous-Paleogene Boundary in the Zaysan Basin, Eastern Kazakstan." *Journal of Stratigraphy* 36 (2012): 417–435.

Mikhailov, K. E., E. S. Bray, and K. F. Hirsch. "Parataxonomy of Fossil Egg Remains (Veterovata): Basic Principles and Applications." *Journal of Vertebrate Paleontology* 16, no. 4 (1996): 763–769.

Mourim, T., P. Bengtson, M. Bonhomme, E. Buge, H. Cappetta, J.Y. Crochet, M. Feist, K. F. Hirsch, E. Jaillard, G. Laubacher, J. P. LeFranc, M. Moullade, C. Noblet, D. Pons, J. Rey, B. Sige, Y. Tambareau, and P. Taquet. "The Upper Cretaceous–Lower Tertiary Marine to Continental Transition in the Bagua Basin, Northern Peru." *Newsletter on Stratigraphy* 19, no. 3 (1988): 143–177. BerlinStuttgart.

Packard, M. J., L. K. Burns, K. F. Hirsch, and G. C. Packard. "Structure of Shells of Eggs of *Callisaurus draconoides* (Reptilia, Squamata, Iguanidae)." *Zoological Journal of the Linnean Society* 75 (1982): 297–316.

Packard, M. J., and K. F. Hirsch. "Scanning Electron Microscopy of Eggshells of Contemporary Reptiles." *Scanning Election Microscopy* 4 (1986): 1581–1590.

Packard, M. J., and K. F. Hirsch. "Structure of Shells from Eggs of the Geckos *Gekko gecko* and *Phelsuma madagascarensis*." *Canadian Journal of Zoology* 67 (1989): 746–758.

Packard, M. J., K. F. Hirsch, and J. B. Iverson. "Structure of Shells from Eggs of Kinosternid Turtles." *Journal of Morphology*, 181 (1984): 9–20.

Packard, M. J., K. F. Hirsch, and V. B. MeyerRochow. "Structure of the Shell from Eggs of the Tuatara, *Sphenodon punctalus*." *Journal of Morphology* 174 (1982): 197–205.

Packard, M. J., K. F. Hirsch, G. C. Packard, J. D. Miller, and M. E. Jones, "Structure of Shells from Eggs of the Australian Lizard *Amphibolurus barbatus*." *Canadian Journal of Zoology* 69 (1991): 303–310.

Scheetz, R. D., and K. F. Hirsch. "SoftShelled Eggs from the Upper Jurassic Morrison Formation of Colorado." *Journal of Vertebrate Paleontology* 15, suppl. 3 (1995): 52A.

VianeyLiaud, M., K. F. Hirsch, A. Sahni, and B. Sige. "Late Cretaceous Eggshells and Their Relationships with Laurasian and Eastern Gondwanian Material." *Geobios* 30, no. 1 (1997): 75–90.

Zelenitsky, D. K., and K. F. Hirsch. "Fossil Eggs: Identification and Classification." In *Dinofest International Proceedings*, edited by D. L. Wolberg, E. Stump, and G. D. Rosenburg, 279–286. Philadelphia: Academy of Natural Sciences, 1997.

APPENDIX B

Announcement of the Honorary Doctorate of Humane Letters for Karl F. Hirsch University of Colorado Commencement, August 11, 1990

A noted researcher and self-taught paleontologist, Karl F. Hirsch is among the handful of experts on fossil eggs worldwide.

By applying modern technology to research on dinosaur eggshells, he greatly advanced the field of paleontology. The study of eggshells, once regarded as museum curiosities, is now recognized as an important sub-discipline within vertebrate paleontology.

Although Mr. Hirsch's formal education in this field was derived from just two University of Colorado classes, his talents have led him to become a respected research associate within the CU Museum and the Denver Museum of Natural History. He collaborates with scientists throughout the world who contact him to identify their findings.

Mr. Hirsch became "hooked" on dinosaurs' eggs in 1974 after stumbling across a prehistoric duck egg while walking through Nebraska's Toadstool Park. Since then he has helped to research and document the discoveries of some of the oldest recorded dinosaur eggs in over 20 scientific journal articles.

Earlier this year he published in the prestigious journal *Science*, research results involving a rare intact fossil dinosaur egg discovered in eastern Utah. The first ever recorded from a 100-million-year gap in the geologic record has been dated at about 145 million years old.

Mr. Hirsch has also been an educational advocate in his field among children, teachers, and nonscientists. He has lectured at various schools, clubs, and organizations, frequently donating fossils to schools and museums across the country.

In recognition of his scientific, scholarly, and educational contributions around the world and at this university, the Board of Regents is pleased to award the degree of Doctor of Humane Letters, *honoris causa* to Karl F. Hirsch.

APPENDIX C

Karl Hirsch's Response upon Receipt of the Harrell L. Strimple Award from the Paleontological Society, 1990

Ladies and gentlemen. Dear members and guests of the Paleontological Society. I am deeply honored to receive the Strimple Award, and I thank everyone who helped bestow this honor upon me and even more, I thank the ones who knowingly or unknowingly awakened the love for fossils in me, stimulated the desire to find out more about these fossils, and the ones who helped in many ways along the road towards this day. I think all these dear people are as important as the one who is receiving the Strimple Award today.

Going back in time, I honestly can say that the thought to become a specialist in one of the fields of paleontology never crossed my mind. I grew up in Berlin, Germany. The fossils in the museum were interesting to look at, but that was about all. After the war my wife and I decided to start a new life by finding a homeland. A church at Port Allegeny, Pennsylvania gave us a new chance at life. After two years we moved to Ohio. Seven years later we took roots in Denver, Colorado, because we loved the outdoors, skiing, camping, and hiking. We could not help but find agates, minerals, and fossils. It was exciting and we became active rockhounds. However, these mysterious remnants of prehistoric life fascinated us more and more and changed the way of our lives.

James A. Garfield once said, "Things do not turn up in this world unless someone turns them up." And we did turn things up and had an ever-increasing desire to find out more about all of them. A talk by Dr. Bill Cobban of the USGS instilled in me a love for ammonites, and he became my guide into this world. My wife had a bigger heart. She loved many kinds of fossils: crinoid, echinoids, trilobites, insects, crabs etc.—in my eyes, too much to handle.

An introductory course on fossils taught by Jordan Sawdo from the Colorado Mineral Society provided my first basic knowledge. When my wife and myself started finding vertebrates, I decided to take two semesters of vertebrate paleontology at the University of Colorado in Boulder. And here I met my mentor, Judith Harris, who really opened the fossil world for me—in short, got me hooked.

In 1973 I reached another important milestone in my life. I found an egg in the Badlands of Nebraska. It was lying there; you could not miss it. It was like someone put it there for me. I wondered what kind of egg it was. Conflicting information and the advice of Dr. Wetmore—"There is not too much done on fossil eggs; if you want to know more you have to find out for yourself." I started my work on eggs. He also gave me the addresses of Prof. Erben from Germany and Dr. Jensen from Utah plus some reference regarding egg literature.

That was 17 years ago. I still love my ammonites but have no time for them: the eggs took over. I examined fossil eggshells and compared them to modern ones. I worked hard. I had to get acquainted with many disciplines that were new for me. I found out that you never finish learning. I also found out that the more you know, the better you know how little you know. However, I loved it, and enjoyed every moment of it.

It was a lonesome way: very few people study fossil eggshell. I corresponded with all of them and met some of them. Through Cindy Carey (UC, EPO Biology) who worked on eggs laid by birds at high altitude, I met Mary Packard (CSU Biology) who then worked on recent turtle eggs. We became good friends and jointly published several papers on modern eggshell. I learned that zoologists look at things differently than paleontologists. I learned a lot from her. It was, however, relatively easy to describe fossil eggshell structure comparable to modern avian, turtle, crocodilian and gecko eggs.

Now I am working with the unknown, the structure of dinosaur eggshell, but here too I found a partner, Konstantin Mikhailov from the Academy of Science in Moscow. Although we have different scientific backgrounds, we have learned much from each other. Hopefully, we will be able to establish a final classification scheme for eggshell structure.

All this sounds like a big story, the story of the self-made man, but it is not so. It is true that you have to be persistent: you have to have a deep desire to find out, and you have to grab opportunities by the neck as they come along. However, I would never have made it without all the people that supported, advised and guided me, the people who taught me the techniques and opened new fields.

Many such people from different institutions, zoos, etc., helped along the road, too many to name them all. The ones who looked down critically at the old rockhound were few; they can be counted on one hand. I think a certain potential lies dormant in every rockhound, if awakened by the professional, it may produce things which will astonish both the professional and the amateur.

I wish to close with heartfelt thanks to all of you who helped elevate me to this important moment in my life; without you I would never have made it.

Reprinted in *Trilobite Tales*, a publication of the Western Interior Paleontological Society, pp. 4–5 (March 1997).

APPENDIX D

Fossils Named in Honor of Karl F. Hirsch

As of the time of the writing of this biography, we are aware that four egg species, "oospecies," have been named in honor of Karl Hirsch. These are as follows, with the relevant scientific article citations. Three represent dinosaurs and one represents a turtle. We are indebted to Cindy Smith of the Western Interior Paleontological Society (WIPS) for providing this information.

Prismatoolithus hirschi

Jackson, F. D., and D. J. Varricchio. "Fossil Eggs and Eggshell from the Lowermost Two Medicine Formation of Western Montana, Sevenmile Hill Locality." *Journal of Vertebrate Paleontology* 30, no. 4 (2010): 1142–1156.

Testudoolithus hirschi

Kohring, R. 1999. "Strukturen, Biostratinomie und systematische und phylogenetische Relevanz von Eischalen amnioter Wirbeltiere" [Structures, Biostratinomy, Systematic and Phylogenetic Relevance of Eggshells from Amniotic Vertebrates]. *CFS Courier des Forschungsinstituts Senckenberg* 210: 1–311. Frankfurt/Main.

Reticuloolithus hirschi

Zelenitsky, D. K., and Sloboda, W. J. "Eggshells." In *Dinosaur Provincial Park: A Spectacular Ancient Ecosystem Revealed*, edited by P. J. Currie and E. B. Koppelhus, 398–404. Bloomington: Indiana University Press, 2005.

Spongioolithus hirschi

Bray, E. S. "Eggs and Eggshell from the Upper Cretaceous North Horn Formation, Central Utah." In *Vertebrate Paleontology in Utah*, Utah Geological Survey Miscellaneous Publication 99, edited by David D. Gillette, 361–375. Salt Lake City: Utah Geological Survey, 1999. https://ugspub.nr.utah.gov/publications/misc_pubs/mp-99-1.pdf.

INDEX

Page numbers appearing in italics refer to illustrations.

Martin Lockley was Professor Emeritus in the Department of Geography and Environmental Sciences, former director of the Dinosaur Tracks Museum at the University of Colorado Denver, and associate curator at the University of Colorado Museum of Natural History. He is the coauthor of *How Humanity Came into Being: The Evolution of Consciousness* and *Dinosaur Tracks and Other Fossil Footprints of Europe*.

Bernie Spilka is a Professor Emeritus in the Department of Psychology at the University of Denver. He is the coauthor of *The Psychology of Religion: An Empirical Approach*, 5th edition.

FOR INDIANA UNIVERSITY PRESS

Tony Brewer *Artist and Book Designer*
Gary Dunham *Acquisitions Editor and Director*
Brenna Hosman *Production Coordinator*
Katie Huggins *Production Manager*
David Miller *Lead Project Manager/Editor*
Dan Pyle *Online Publishing Manager*
Pamela Rude *Senior Artist and Book Designer*
Stephen Williams *Marketing and Publicity Manager*